With awe of, and gratitude to, Hashem.
HaShem means The Name, aka The One designer and creator, aka G-d of Abraham, just as described by Moses.

'Inquire to the ends of the universe...'[1]
Consider astronomy to proclaim the handiwork of Hashem.[2]
'Hashem said let there be light...'[3]
Understand the empirical observations via awe of Hashem and selfless Torah study...[4]
'Truth springs from Earth and just measure... from above...'[5]
'..as signs for the .. and years.'[6]

     Baruch Hashem, the Haskamah - approbation is from Rav Rachamim Pauli, SHLITA, whose physics royalty 'Yichus' includes great Uncle Wolfgang Pauli.
     May we, please Hashem, help advance the Torah and science far beyond the marvelous achievements of Albert Einstein & Wolfgang Pauli. Resulting in the true description of our empirical observations.

---

[1] Deuteronomy 4:32
[2] Tractate Shabbos 75a Rav Zutra b. Tuvia.
Isaiah 5:12, 40:26,
Psalm 19:2 & 148:1-6 by 'Kiddush Levanah': '.. and He set them up to eternity, yea forever, He issued a decree, which will not change.'
www.chabad.org/library/bible_cdo/aid/16369
[3] Genesis 1:3..
   T' Chagiga 12a
   Mishlei 13:9 light of the righteous..
   Isaiah 60:19-20 light independent of sun &r moon..
[4] Psalm 25:14, Mishna Avot 6:1
[5] Psalm 85:12
[6] Gen. 1:14

ב"ה

Haskama

I have read a number of books written by Orthodox Scientists on Creation. That of Prof. G. Schroeder, Prof. Aviezer, hear a synopsis of my former Chevruta Prof. Lee Spector. Each of the previous books deal with a religious – modern scientific match. Professor Schroeder uses Kabbalah to get to the 13,800,000,000 year-old universe.

All the latest modern Cosmology models, have a number of flaws. First, they need mysterious dark energy which they cannot explain. Second, they cannot explain the current state of the Universe which requires massive movement of the material for forming early star clusters faster than the speed of light.

Now who is to say that initial Big Bang of the whole material at a temperature of 10 to the 42nd power Kelvin degrees in an 11 dimensional universe would behave according to Einstein's theory due to the heat, the small particles in dimensions above time, the 4th dimension?

The only reasonable theory that has no need for this mysterious dark energy is the Pearlman: Stars Preceded Inflation - Redshift Attests to Lagging Light Theory. The late Lubavitcher Rebbe, who was an engineer and had learned all the scientific theories, contended that the Universe was created in six days of twenty-four hours each. He also knew Kabbalah better than Professor Schroeder.

The Pearlman Theory explains better than the standard model the current observable Universe after the Big Bang as no dark matter is necessary.

Sincerely,

Rabbi (Richard) Rachamim Pauli

Administrator Rosh Kollel and Director of Ethics Kollel Beit Shlomo Kiriat Sefer, Israel

Former Student Assistant on the Cosmic Ray Project of City College New York Physics Dept.

Former Member of the Prairie Meteor Project of the Smithsonian Astrophysical Observatory

Address;   כתו
5 Or Hachaim apt.2   רח' אור החיים 5/2
Kiriat Sefer, Modiin Ilit Israel 71919   קרית ספר, מודיעין עילית

2

Haskama – Kollel Beit Shlomo

I have read a number of books written by Orthodox Scientists on Creation. That of Prof. G. Schroeder, Prof. Aviezer, hear a synopsis of my former Chevruta Prof. Lee Spector. Each of the previous books deal with a religious - modern scientific match. Professor Schroeder uses Kabbalah to get to the 13,800,000,000 year-old universe.

All the latest modern Cosmology models, have a number of flaws. First, they need mysterious dark energy which they cannot explain. Second, they cannot explain the current state of the Universe which requires massive movement of the material for forming early star clusters faster than the speed of light.

Now who is to say that initial Big Bang of the whole material at a temperature of 10 to the 42nd power Kelvin degrees in an 11 dimensional universe would behave according to Einstein's theory due to the heat, the small particles in dimensions above time, the 41h dimension?

The only reasonable theory that has no need for this mysterious dark energy is the Pearlman: Stars Preceded Inflation - Redshift Attests to Lagging Light Theory. The late Lubavitcher Rebbe, who was an engineer and had learned all the scientific theories, contended that the Universe was created in six days of twenty-four hours each. He also knew Kabbalah better than Professor Schroeder.

The Pearlman Theory explains better than the standard model the current observable Universe after the Big Bang as no dark matter is necessary.

Sincerely,

Rabbi (Richard) Rachamim Pauli

Administrator Rosh Kollel and Director of Ethics Kollel Beit Shlomo, Kiriat Sefer, Israel. Former Student Assistant on the Cosmic Ray Project of City College, New York, Physics Dept. Former Member of the Prairie Meteor Project of the Smithsonian Astrophysical Observatory

Title Page:
The 'Moshe Emes' series Vol. II for Torah and Science Alignment
Pearlman YeC 'SPIRAL' cosmological model (2013)
File: 29 Adar 5786
Last edit March 18, 2026
SPIRAL= Stars Preceded Inflation, their Redshifted Light Attests.
SPI = hyper-dense proto-galactic (Stellar) formation Preceded
nearly all cosmic-Inflation expansion.

If SPIRAL one would predict the vast body of empirical
observations that is the prevalent cosmological redshift (CR) of
distant starlight and the overall increase of that CR w/distance.

SPIRAL's 'cosmological blue-shift offset' hypothesis helps
explain why an overall increase, but not straight-line increase, of
CR with distance.

RAL = Cosmological Redshift of their Light-trails Attests to those
stellar objects having receded from us during, & when subjected
to, cosmic inflation expansion.

We also explain why the Years elapsed subsequent to the
hyper-dense start that ended in gravitational bound equilibrium is
= to the Light Year (LY) distance of the nearest departure point of
any light arriving here and now at standard light speed, ever
subjected to cosmic expansion.

Dedications:
With gratitude to those named below,
please G-d for an Aliyas Nishamah (elevation of their soul),

Name followed by Yahrzeit date:

Teviah Velvel 'Thomas Wolfe' b. Yitzchak Pearlman   Nissan 29
Miriam (Pearlman) bas Akiva (& Pearl Leah Vogel) Cheshvan **13**
Yitzchak 'Israel' ben Abraham & Anna Pearlman   Kislev 14
Chayah Esther 'Ida' Pearlman bas Moshe Aryeh & Itah Sarah Iyar 05
Moshe Aryeh b. Mordechai Mines dad of Ida   Nissan 11
Itah Sarah 'Anna' (Mines) bas Tuviah Levine Ida's mom Iyar 29
Abraham b. Meir Pearlman dad of Israel   Shevat 17
Chana Esther 'Anna' bas Wolfe mother of Israel   Adar 18
Akiva 'Charles' ben Pinchas ben Binyamin Vogel   Cheshvan 02
Pearl Leah 'Pauline' bas David & Rebecca Hacker   Av 09
Pichas b. Zev Dov 'Beryl/Benjamin' dad of (Akiva Vogel) Kislev 14
Soesie Breena (Vogel) bas Tzvi Zev (Hirshon) mom of //) Cheshvan 20
David b. Abraham HaLevi Hacker dad of Pearl Leah   Adar I, 03
Rebecca bas Mordechai (Bloch) Hacker mom of Pearl Leah Shevat 13
Yehoshua Pinchas 'Josh' b. Teviah Velvel & Miriam Nissan 08
Chana Binah 'Ann Brenda' bas Teviah Velvel & Miriam Cheshvan 16

Introduction:

After its introduction in 2013, SPIRAL has been holding up to scrutiny and is panning out as the highest probability explanation of all the empirical cosmological observations. Based on basic physics and math.

This book was a discovery process in the quest to reconcile a light speed limit of standard light speed within Mosaic Testimony. BH': Mission accomplished!

Einstein and Galileo are on record with statements in line with: 'One layman with a higher probability hypothesis, explaining the empirical observations (definition of the stronger science), outweighs an unlimited number of credentialed scientists, that still hold by the competing consensus champion theory'.

SPIRAL is already due to replace the competing SCM-LCDM model as the new standard in cosmology, after fair comparison and dissemination.

Ironically the physics exposes scoffers as projectionists. As it not only aligns best within Abrahamic faith but has now properly falsified consensus science that contradicts Moses.

In appreciation of my family, friends, teachers, colleagues and students.
That all benefit from The Torah Sages, who kept the unbroken chain of Torah Messorah.
All under The Divine Providence of The One Heavenly Father.
It turns out the vastly higher probability science attests Abraham's Faith and Mosaic Testimony were right to begin with.

AM   = Anno Mundi year count (start year one on day six)

B,M,k = B=Billions, M=Millions, k not K = Thousands

'c'   = Light Speed. The standard speed of light.

CB   = Cosmological Blue-shift offset

CE   = Cosmic (metric) Expansion of space (including CI)

CI   = Hyper Cosmic Expansion (if SPIRAL all CE)

CMB = Cosmic Microwave Background Radiation

Cosmological Principle: universe on large scales homogeneous & Isotropic.

Copernican Principle: our location in the universe isn't privileged.

CR   = Cosmological Redshift (due to CE)

EMR= Electromagnetic Radiation. GRB=Gamma Ray Burst

IU = Inner Universe Visible Light from stellar objects therein was never subjected to Cosmic Expansion (CE) so not CR'ed or CB'ed. LY radius = to year count post CI.

LCDM = Lamda Cold Dark Matter w/Dark Energy & Reg. Matter

LY   = Light Year/s. Distance light travels per year at 'c'.

NASA = National Aeronautics and Space Administration

OCE   = Ongoing Cosmic Expansion / Hubble Flow

OU= Outer Universe Starlight therefrom was subjected to CE, departing when universe still compressed, prior to mature density.

SCM   = Standard Cosmological Model currently LCDM.

SSSO = Stable Steady State Oscillation

STR = Einstein's Special Theory of Relativity

Visible (observable) Universe = Sphere of detectable matter.

YA = Years Ago   YeC = Young Earth Creation

SPIRAL = SPI-RALL predicts CR and it's increasing w/distance.

SPI = hyper-dense proto-galactic (Stellar) formation Preceded Cosmic-Inflation expansion.

RA = Cosmological Redshift Attests to their (those stellar objects receding from us during cosmic inflation).

LL = Lagging Light (decreased frequency/increased wavelength) as that distancing extends their arrival timespan.

"I want to know God's thoughts," said Einstein."... the rest is detail." "Distant Starlight and the Age, Formation, and Structure of the Universe" will give you some of those details! Welcome to SPIRALL: "Stars Preceded Inflation - Redshift Attests to Lagging Light."

How excellent to read of the ins and outs of God's thoughts on the creation of the stars. Pearlman's SPIRALL hypothesis is the best explanation for what we observe in space as it correlates with divine revelation in the Torah. Indeed, the author compares SPIRALL with 29 points of the Standard Cosmology Model. You be the judge of which model better explains the formation of the universe.

Have pen and notepaper on hand! You're stepping out to the cosmos -- new terms and abbreviations must be used. We're not in Kansas any more! Possibly the End of the Standard Cosmology Model"

'SPIRAL' cosmological redshift hypothesis and model
File: 29 Adar 5786   www.amazon.com/dp/B09L3NP43P
Article list from July 07, 2023 until Jan. 19, 2026
Cite: Roger M. Pearlman YeC   CC BY 4.0
Title / Description   DOI-Link   Month / Year

New Cirinus Black Hole data in light of SPIRAL. January 2026
DOI:10.13140/RG.2.2.10171.07207
The First 96 Hours - Size and Age of The Universe based on
CMB Temperature. DOI:10.13140/RG.2.2.36546.44482 Dec. '25
Radio Dipole Anomaly resolved! by SPIRAL cosmological
redshift hypothesis and model on the cosmic distance ladder.
DOI:10.13140/RG.2.2.16275.34080   November 2025
Quasar Dipole and CMB in light of Pearlman SPIRAL
DOI:10.13140/RG.2.2.10806.36165   November 2025
Pearlman Cosmology hyper-dense proto-galactic formation helps
explain the age and structure of the universe.
DOI:10.13140/RG.2.2.21890.18881   .May 2025
SPIRAL universe size at decoupling CMB calibrated.pdf
DOI:10.13140/RG.2.2.29656.20484   April 2025
Pearlman first and best explanation of Oxygen in the earliest light
trails b.pdf DOI:10.13140/RG.2.2.26144.49925 April 2025
Feynman's Doubt DOI:10.13140/RG.2.2.35508.36484 Dec. 2024
CMB distribution not 'being 'checkerboard' is explained best by
Pearlman SPIRAL cosmology and discuss the gravitationally
bound region relation to SPIRAL light year radius i.
DOI:10.13140/RG.2.2.29855.98721   August 2024
Pearlman on 'Rakiah' in scripture - Metric Expansion of Space.
DOI:10.13140/RG.2.2.30240.49923   March 2025
SPIRAL: No Fudge Cosmology   July 2024
10.13140/RG.2.2.32319.52641 & 10.13140/RG.2.2.13069.96480
SPIRAL cosmological model 'MVP' Hypothesis   June 2024
DOI: 10.13140/RG.2.2.14477.35041
JWST Rapid Mass Assembly and Metal Enrichment findings
confirmation of Pearlman YeC SPIRAL cosmological redshift
hypothesis and model   June 2024
10.13140/RG.2.2.20373.13287 & 10.13140/RG.2.2.24054.87367

Table of Contents:

---

[7] Einstein's Doubt, Biggest Blunder and Regret Oct.'20 DOI:10.13140/RG.2.2.14194.72648

[8] Parallax, Gravity and The Cosmic Distance Ladder April '16 10.13140/RG.2.2.12752.93449

[9] 'GRIP' Galactic Rotation without Dark Matter Oct. '20 :10.13140/RG.2.2.11678.14404

[10] Olbers'_Paradox_Pearlman_YeC_evaluator June '23 DOI :10.13140/RG.2.2.29530.72648

[11] SPIRAL's Magnetic Repulsion file 7.46b pdf July '20 DOI:10.13140/RG.2.2.11829.13281

[12] 'Blue-shift Offset' (to Cosmological Red-shift) hypothesis part of 'The Pearlman SPIRAL' cosmological red-shift hypothesis. May 2016 DOI :10.13140/RG.2.1.2072.9205

[13] SPIRAL cosmological model's 'Black-hole Illusion Resolution at Galactic Centers' January 2018 DOI :10.13140/RG.2.2.31968.08966

[14] Pearlman YeC SPIRAL 'Jiffy Pop' 'Electro-'Magnetic Repulsion' on Cosmic Inflation Expansion Jan. 2018 DOI:10.13140/RG.2.2.11914.07364

[15] Pearlman 'GRaB' hypothesis March 2022 DOI :10.13140/RG.2.2.23153.75361

[16] Distant Starlight with CREST, CMB comments April'16 :10.13140/RG.2.2.19378.88002

[17]Pearlman YeC SPIRAL 'CoMBO' Hypothesis Dec. '21 DOI:10.13140/RG.2.2.30985.04962

[18]SPIRAL vs SCM relative dating based on CMB Aug. '22 : 10.13140/RG.2.2.15438.93760

[19]'MVP' hypothesis .. exhibit A   January 2015 DOI: 10.13140/RG.2.2.33824.99846

[20]'Draw Play Lunar Formation Hypothesis.. April '17 DOI:10.13140/RG.2.2.11075.21289
   SPIRAL's 'Draw Play' info-graphic   June 2017 DOI:10.13140/RG.2.2.15200.07687

[21]'SNAP' Proto-Earth origins hypothesis January 2018 DOI:10.13140/RG.2.2.20224.03841

[22]'Pearlman vs Hubble'   July 2018 DOI:10.13140/RG.2.2.23703.59043

[23]Dayenu Fistful Nobel's Worth scientific advances Aug.'20 10.13140/RG.2.2.10852.39044

[24]Pearlman on the Keating Checklist   Aug. '22 DOI:10.13140/RG.2.2.11676.91528
   Pearlman Lab to test SPIRAL..   May '21 DOI:10.13140/RG.2.2.11787.50728

[25]'SPIRAL 'SAFETY'   Oct. '21 DOI:10.13140/RG.2.2.13061.27367

[26]Pearlman SPIRAL 'SOD'   Jan. 2022 DOI:10.13140/RG.2.2.36817.75360

[27]SPIRAL vs SCM-LCDM on Tolman Test Oct. 2022 DOI:10.13140/RG.2.2.12530.11201

[28]JADES-JWST Lyman-break   Jan. '23 DOI:10.13140/RG.2.2.20079.85924
   and JWST helps corroborate SPIRAL Aug. '22 DOI:10.13140/RG.2.2.15840.56322

[29]SPIRAL hyper-dense proto-galactic formation preceded 'hyper cosmic-expansion day four'
   SPIRAL update 'Cosmic Inflation, Day Four' Aug. '22 DOI:10.13140/RG.2.2.13354.63687

[30]Mankind is The Focal Point of The Universe Aug. '22 DOI:10.13140/RG.2.2.16290.64963
   We are special. SETI Sacked.pdf   Dec. '22 DOI:10.13140/RG.2.2.16675.31520

'Hashem gave them (sun, moon, stars) into the heavens to give light upon the Earth.' [31]
'Hashem made the Heavens and Earth in full Stature' [32]

'Distant Starlight and the Age, Formation and Structure of the Universe' by Roger M. Pearlman

The Pearlman SPIRAL cosmological redshift (CR) hypothesis and cosmology model was formulated by April of 2013 (+3760=5773 Anno-Mundi (AM) when start year one on day six).

SPIRAL is an internally consistent model that predicts CR rather than react to it. The basic idea has held up well to subsequent factual observations and peer-review scrutiny. If the actuality, it's where Torah & Science align.

SPIRAL is an alternate 'Big Bang' cosmology model that is consistent with, a start, by a singularity, then Cosmic Expansion (CE), but not ongoing cosmic expansion (OCE). CE lowers light oscillation frequency, so lengthens the wavelength, resulting in cosmological redshift (CR).

SPIRAL helps revise the how, where and when of stellar formation, so enhances our understanding of the age, formation and structure of the universe.

SPIRAL makes the case for a start, by a hyper-dense minuscule sized universe, with proto-galactic formation prior to hyper expansion, ending in a stable, on the first plateau steady state oscillation universe, of mature, modern, size and density, relatively early in history. The Earth-Sun ecliptic is the approximate center of the sphere of the 'visible universe', which approximates the entire universe.

SPIRAL's 'MVP' hypothesis shows why ours is the vastly preferred view in the entire universe.

---

[31] Genesis 1:17 Part of the day four CI Inflation Epoch.
[32] Gen. 2:1 see Talmud Chullin 60a & T' Rosh Hashanah 11a

Faulty premise and assumptions are a leading cause of the confirmation bias that permeates current popular science. With no ongoing cosmic expansion (OCE) and us by the center SPIRAL better explains all the factual observations.

If cosmological redshift (CR) isn't due to OCE but attests we're by the approx. center of the universe, it falsifies The Copernican and Cosmological Principles, premise of modern cosmology, that assume 'we occupy no special position in the universe, it's Homogeneous & Isotropic..'[33]

So when we show premise assumptions are low probability at best, it's an eye opener for pure scientists. Resulting in new insight and advancing the science.

We provide a way to reconcile distant starlight with The Torah narrative and timeline. We identify empirical evidence for Young Earth Creation (YeC).

We illustrate why we have the optimal view-point of the universe. SPIRAL provides a stronger solution to Olbers' paradox than The Standard Cosmological Model (SCM).

SPIRAL helps resolve open questions in science like the galactic rotation problem, large structures that challenge a homogeneous and isotropic distribution of matter, and the CR of distant starlight, without dark energy and dark matter.

If SPIRAL is the highest probable explanation of the natural observations, it should replace SCM-LCDM as the new scientific standard. How much more so if SCM is falsified and SPIRAL the last credible hypothesis standing.

---

[33] www.astro.ucla.edu/~wright/cosmo_02.htm
Copernicus who held by a solar centric universe, good name was hijacked!

SPIRAL provides a more accurate cosmic distance ladder. The natural observations in light of SPIRAL add up to a smaller, more parsimonious, and younger universe than if SCM.

If valid the overwhelming empirical evidence like the prevalent CR of distant starlight falsifies / invalidates all deep-time dependent scientific hypotheses and assumptions.

The stronger science is the higher probability of the explanation of the natural observations. Always use available context. For the truth to rise to the top, ensure study and fair consideration of alternatives that may include the actuality.

Incredulity, lack of fair consideration and equating consensus science with the strongest science, is anti-science and a recipe for stagnation. Science isn't a popularity contest.

The greatest advances in science are apt to start with a minority of one, outside the consensus box, and be met with incredulity. If not someone else should have the same idea. So don't be blinded by deep-time dependent doctrine (DTD) dogma. 'All I am saying is give science (+ peace) a chance' :)

Let us start with a premise there was only one historic actuality to get us to this point. So while there may be trillions of ways that could not get us to this point, there is at least one way that could. Perhaps only one way.

Is The Current Standard Cosmological Model (SCM) even internally consistent and able to align within all the empirical observations? As we found SPIRAL is and does. Test and grow.

Before considering the positions and the evidence science can't exclude either. If one is true the other is not. With current data we may rule one out, based on basic physics and math.

The Moshe Emes 'RCCF' Framework six principles to understand science, helps explain how/why independent dating methods, including radiometric dating, favor YeC.[34]

SCM is cumulative. For example, biased on radiometric rock dating assumptions to date the Earth then uses that to cross calibrate a DTD age for the universe.[35]

---

[34]Cites ICR rate team great John R. Baumgardner & Yaacov Hanoka PhD
[35] 'Cosmology & Particle Astrophysics' by Lars Bergström, Ariel Goobar

'HTP' hypothesis: Early & Rapid Element & Galactic Formation

SPIRAL's 'HTP' hypothesis: In RCCF we reference how heat, time and pressure (HTP) are three variables in the rate of rock formation. An increase of heat, and or pressure, can reduce the time. Think of how quickly industrial diamonds are made, even faster if with the heat and pressure of an asteroid impact.

The same variables apply in galactic & stellar formation. Think of all the concentrated matter & energy present before the inflationary expansion epoch. This touches on how a Hyper Dense proto-galaxies and Stars Preceded cosmic-Inflation expansion event the 'SPI' in SPI-RAL. The current consensus alternative is speculative at best.[36]

If Trillions of galaxies means The One designer creator aka G-d of Abraham had trillions of times as much for the natural mechanism (the forces of nature and laws of science themselves being part of the design/creation) during the hyper-dense proto-galactic formation phase when all in an area less than 1 galaxy.[37]

Be the actuality gravity, electro-magnetic radiation, strong nuclear force, weak nuclear force, vibration, and/or a combination and/or hybrid. you had it. At hand was sufficient for near instantaneous galactic formation, which per SPIRAL could've been as early as day one and certainly by the day four cosmic inflation expansion epoch as by the end of that event the universe was at it's mature density with no subsequent cosmic expansion.

Based on CMB temperature SPIRAL primordial gas cloud, element and proto-galactic formation was 2.5M times earlier, and more rapid ie when the universe smaller than if SCM, where already relatively very early and rapid.[38]

Per Rashi start proto-galactic formation mid-day one, and done by hour one day 4 per Pirkei D'Rebbi Elizer (7).

Each hyper-dense proto galaxy relates to the formation of it's electromagnetic field. That relate to the CI expansion epoch. So, said fields approximate the number of galaxies.

---

[36]www.icr.org/article/nebular-hypothesis-doesnt-hold-together Jack Hebert PhD 2018
[37]https://phys.org/news/2017-01-universe-trillion-galaxies.html
   T' Brachos 32b 12X30(a)x30(b)x30(c)x30(f)x365,000 stellar enumeration.
[38] Latif, M.A.,et al. Turbulent cold flows gave birth to the first quasars. Nature 2022
If SPIRAL entire univere area = to 1B, vs 46.5B, LY radius sphere, times 200, if SCM.

An Earth-centric universe fits well with scripture. SPI-RAL is how/why an Earth-Sun ecliptic centric spherical universe is a high probability science explanation of the observations. The approximate center of the universe, not of our solar system...

An Earth-Sun ecliptic centric universe, our optimal view, no ongoing cosmic expansion (OCE), a universe thousands, not billions, of years old, a more accurate cosmic distance ladder, & a precise way to date the age of the universe, are derived by SPIRAL understanding of cosmological redshift (CR).

By providing a stronger premise one can upgrade an entire field of science as we do here by CR not due to OCE.

CR due to OCE is deep-time dependent. If the actuality is otherwise, SCM will never align with all facts.

Credentialed YeC scientists have a big advantage over those who rely on DTD hypotheses. As scientists who advocate YeC had to learn DTD to get credentialed. Consensus DTD biased explanations have some facts like a hyper-dense start but miss the context to fully understand them. If SPI one would predict CR and it's overall increase with distance, due to past, not ongoing, cosmic expansion. So, no OCE required, yet missing, dark energy.

As of 5779 anno-mundi (AM), light visible here and now from stellar objects 5779 Light Years (LY) and greater distance, departed when 5779 LY distant, 5779 years ago.[39]

Early stellar formation attests to Intelligent Design in complex stature. The observations don't show subsequent ongoing stellar formation, nor evidence of deep-time.[40]

If SCM conditions a trillion years mightn't be enough for stellar and heavy element formation. If SPIRAL hyper-dense proto-stellar formation, four days was ample time for all the processes, to get from minuscule to the mature size & density. 4/365.25(SPIRAL LY radius 'I')5783 a fraction into history.

Later we give HTP like examples in nature and on GRB, Quasar, X-Ray Binary, Tidal Disruption Events, Colliding Wind Binary, Black-holes, galactic interaction...

[39] arxiv.org/pdf/1603.00461v1.Oesch, P.A.'A Remarkably Luminous Galaxy'
[40] Craig Gates, Woodland Hills Telescopes, Tehachapi BVS Observatory

Three Pillars of The Big Bang
'..the three cornerstones of The Big Bang model are:

One: The blackbody nature of the CMB spectrum.
Two: Red-shifting of distant galaxies for uniform expansion.
Three: The observed abundances of light elements (in particular helium and heavy hydrogen), indicating that they were "cooked" throughout the Universe at early times.[41]

All three of these pillars for SCM can align with and support SPIRAL too. So SPIRAL deserves = consideration.

SPIRAL & SCM 'big bang' component indicate the physical universe, time, and laws of nature, had a start.

Cosmic Inflation Theory is an integral part of SCM. Cosmic Microwave Background Radiation (CMB) indicates the physical universe expanded at speeds of at least 100X light-speed. Hyper cosmic expansion fits SPIRAL as well.

If SPIRAL 'cooked throughout' while still hyper-dense.

SPIRAL reconciles YeC with starlight from distant stars and Einstein's Special Theory of Relativity (STR) which limits light speed to the standard speed of light.

If valid SPIRAL caps years elapsed subsequent to the cosmic inflation expansion event to the number of years = to the LY distance to the nearest departure point of any light that's been subjected to cosmic expansion. So millions of years have not elapsed. So the prevalent cosmological redshift of distant starlight and STR falsify all deep-time dependent scientific hypotheses, and attest to YeC.

Prior to formulating SPIRAL, time dilation & the role of the observer, by Prof. Moshe Carmeli, seemed the best fit. Dr. Alexander Polorak, Dr. Gerald Schroeder and some other creation science astrophysicists held likewise.[42]

---

[41] Douglas Scott PhD www.astro.ubc.ca/people/scott/faq_basic.html

[42] Professor Moshe Carmeli, The First Six Days of the Universe. B'or Hatorah Journal # 15. 2005 & arXiv:astro-ph/0103008

Some scientists concur with the SPIRAL position of no Ongoing Cosmic Expansion (OCE). Holding CR is not proof of OCE and or that other observations preclude OCE.

Doctors: John G. Harnett[43], Thierry De Meese[44], Halton Arp,[45] Hilton Ratcliffe,[46] Michael Peck,[47] Erik J. Lerner,[48] Otto E. Rossler,[49] Yuri Heymann,[50] Robert Gentry,[51] Ezra Segal,[52] and Einstein before capitulating to Hubble.[53]

Dr. John G. Hartnett early peer review scrutiny of SPIRAL concerns regarding CMB and CR were taken into account and answered in the editions that followed, as were concerns of other scientists regarding parallax and more.

Without SPIRAL it is understandable why many who hold there's no ongoing cosmic expansion (OCE), don't hold a big-bang, and why those who do hold big-bang hold OCE.

If CMB and CR align with SPIRAL this hybrid big-bang model should become the new standard. Per SPIRAL Cosmic Expansion (CE) that caused CR (and if SPIRAL's: 'Blue-Shift Offset', some blue-shift too), ended long ago.

---

[43]Is the Universe really expanding? arXiv:1107.2485 [physics.gen-ph] 11/19/11
[44]'The Karlsson Peaks in the Quasar's Redshift Distribution as an Indication for Circling Light in a non-expanding Universe' Thierry De Meese May '15
• [45]"Seeing Red" ISBN 0-9683689-0-5 Halton Arp PhD 1998
[46]'The Static Universe: Exploding the Myth of Cosmic Expansion' Hilton Ratcliffe '10
[47]Evidence of a Global Gravitational Potential Astro. Review V9 7/14
[48]UV surface brightness of galaxies arXiv:1405.0275[astro-ph.CO] '14
[49]eujournal.org/index.php/esj/article/view/2762/2614
[50]ptep-online.com/index_files/2014/PP-38-11.PDF Tired Light
[51]Robert Gentry PhD, http://www.orionfdn.org/papers/index.htm
A New Redshift Interpretation: http://arxiv.org/abs/astro-ph/9806280
Affirmed: http://arxiv.org/abs/physics/9810051 and
The Genuine Cosmic Rosetta: http://arxiv.org/abs/gr-qc/9806061
[52]'Einstein's Static Universe: An Idea Whose Time Has Come Back?' Aubert Daigneault & Arturo Sangalli www.ams.org/notices/200101/fea-daigneault.pdf
[53]'Einstein's Static Universe' arxiv.org/pdf/1203.4513 Domingos Soares

It can be as simple as considering the perspective using the competing school of thought assumptions. Here we address how factual cosmological observations are not only consistent with RCCF-YeC narrative but independently attests to it.

Consider the current Standard Cosmological Model (SCM-lCDM) where the universe starts by a singularity, is 13.8B years old, our 'visible universe' radius is now about 46.5 billion (B) light years (LY) to date. Assume the most distant visible stars formed 4B max LY away, 13.4B years ago and are now 46.5B LY distance from Earth. Their starlight has traveled 13.4B LY, and taken 13.4B years to reach here, due to 9.4B LY distance of CE.

13B rounded gives 800 Million (M) years for stellar formation in the most distant visible galaxies in every direction. Recently detected light estimated from 13.4B LY and years ago leaves only 400M max till stellar formation. 'Earlier than predicted complexity' indicates Intelligent Design and fits within YeC best.

The radius of the sphere that is the visible universe has expanded 11.6 fold 46.5B LY distant now vs the 4B LY distance at stellar formation = 42.5B LY increase. 42/4 = a 10.625 fold increase. The average rate of increase in the radius is about 3.27c(c = Light Speed) =42.5BLY/13B years.

Every ten-fold expansion of the radius is a thousand-fold expansion in the area of the sphere.[54]

Per SCM & SPIRAL the universe started hyper-dense followed relatively soon by rapid cosmic inflation expansion (CI).

When we use the term Ongoing Cosmic Expansion (OCE) in this work we're referring to alleged Hubble Flow expansion of the universe post a Cosmic Inflation event. Cosmic Inflation itself being a type of hyper cosmic expansion.

Under SCM cosmic inflation refers to a event prior to stellar formation. Under SPIRAL cosmic inflation expansion was subsequent to proto-stellar formation.

---

[54]Each doubling/halving of the radius = 8 fold increase/decrease in volume.

Distant starlight has prevalent CR. Overall the CR increases with the stellar objects distance. This helps locate us by the center of the visible universe.

The Hubble Constant describes this rate of increase. That steady increase may argue there was no material pause in CE. Over the course of the past 13B years if SCM[55] During the CI-CE event if SPIRAL.

Under SCM and SPIRAL CR is consistent with CE. Under SCM if there had been no CE during the past 13B LY there should be no CR, only regular light from stars within 13B LY radius from us. No material pause in CR in no way proves OCE. As it's also consistent with SPIRAL's CR not due to OCE, but due to past CE during cosmic inflation (CI).

How could any stars and galaxies form after CI under the SCM assumption of an even distribution of matter, if the universe continued to expand and thus was still moving apart and spreading out a very rapid pace above c?

So if one holds stellar/galaxy formation across deep space largely due to gravity, a pause in CE would help. As gravity has by far the weakest coupling constant of the four fundamental forces.[56]

Either way SCM holds the CI lasted a short time and came to an abrupt end. SPIRAL also holds CI ended. Just that CE did not resume. That after proto-stellar formation we paused into our current sized stable 'big bang into a static' stable steady state oscillation universe.

---

[55]'Observational Evidence from Supernovae for an Accelerating Universe and a Cosmological Constant' Adam G. Riess, 1998 arxiv.org/abs/astro-ph/9805201 and 'Measurements of Omega and Lambda from 42 High-Redshift Supernovae' S. Perlmutter,. 1998 http://arxiv.org/abs/astro-ph/9812133
[56]http://hyperphysics.phy-astr.gsu.edu/hbase/forces/couple.html#c1

Lack of evidence of a post CI pause for stellar formation favors pre-CI proto-stellar formation 'SPI'.

With Cosmic Inflation's Expansion cooling heat, dissipating pressure or other factor, could have removed the catalyst for that expansion, stalling it in equilibrium.

The hyper CI-CE may have even helped distinguish the nature of the fundamental forces into what we are familiar with now.[57]

If a force slowed the CE to a stop why didn't it keep contracting? Why would it start expanding again? Consider issues like 'Inflation Energy' and if energy density was below critical value. If SPIRAL no CI 'miracle' for 'flat' universe.[58]

Perhaps you hypothesize missing dark energy? If it was already present, why did it allow expansion than a pause to begin with? If it formed when the stars did how did it? Can there be accelerating CE with/out it? Why is there no direct empirical evidence of such a massive force? If there is no OCE/contraction what justifies it's existence? [59]

If CE resumed after stellar formation how could any new galaxies form without another pause? If stellar formation was in deep space what kept the matter that formed the galaxies from continuing to contract into one massive star or black hole at each galaxy center?

Starlight emitted during an alleged pause in CE for galaxy and stellar formation should be emitted with no cosmological redshift. See our CREST hypothesis.

CR is prevalent on visible starlight from objects now outside our galaxy and local group. SCM assumes OCE to explain that prevalent CR. If no CR no reason to think OCE.

---

[57]'The First Three Minutes...Of The Origin Of The Universe' Steven Weinberg
[58]The cyclic Universe Paul Steinhardt PhD in 'The Universe' John Brockman
[59]https://science.nasa.gov/astrophysics/focus-areas/what-is-dark-energy

To my knowledge there is no direct evidence in the CR measurements of distant starlight that indicate a material pause for stellar formation in deep space after the initial era of stellar formation.

That makes the conditions for subsequent stellar formation in deep-space even lower than if there had been a additional pauses of OCE.

Whatever caused stellar formation had to have a very high probability of success, as there are so many today. Be the stellar formation nearby (SPIRAL) or spread out (SCM). The observations do not match an uniformitarian formation rate over deep-time or even many different times. Both SCM and SPIRAL put most formed about the same time, early on.

The way to get that high a probability is with a ready, willing and able designer/creator. The design/creation includes laws of nature consistent with the task. A random operating system with an override option is a design choice!

No major imbalance of nearby stars over more distant stars indicates most stars had to have formed early in history.

That most (all?) stars formed in galaxies add another clue for the right conditions and timing required.

Hypotheses behind exactly how stars formed under SCM assumptions include speculation and are uncertain.

It's reasonable to assume there's more to stellar formation than we know yet. With the right conditions stellar formation can be near instantaneous. Think of liquid to gas, lava to rocks, rocks to liquids,.. with heat/pressure...

If Electromagnetic Radiation dominated early, then matter, maybe why SPI left the collimated light trails by CI.

All energy & matter was in proximity prior to CI if SCM or SPIRAL. So all required for stellar-formation on hand. Hyper-density allows maximum pressure so minimal time. Sound carries better in a denser atmosphere! Compare whale calls carry miles in water vs zero on zero atmosphere Moon.

With SPIRAL proto-stellar formation preceded water, but water preceded Cosmic Inflation (CI). Water resistance shortens wavelength and increases frequency. An early offset to CE's oscillation frequency shortening, wavelength lengthening.[60]

If SCM, and regular light became redshifted along the way, if a resumption of Cosmic Expansion (CE) post stellar-formation, it'd take longer to get here to cover the greater distance.

Visible stars formed within 4B LY within 800M years of the formation of the universe and big bang. If the universe is only 13.8B years old. No light that departed from 13B LY away 13B years ago (YA) could reach here now if any ongoing CE (OCE).

That is why under SCM the most distant visible stars whose light we see now formed 4B+/- LY away. Due to CE the light has traveled 13B LY over the course of 13B years. So it has been subject to 9B+/- LY distance of CE. Without any OCE the same light would have passed us 13-4=9B YA.

SCM holds all this time the universe has been expanding and the same stars 4B LY away 13.4B YA are 46.5B LY distance now. While CR is evidence subjected to CE, if OCE we dispute.

The Hubble constant calculates the maximum LY distance visible starlight departed from, if SCM, 13.4B years ago.[61]

SCM may not just hold OCE, but accelerating OCE. The OCE might be linear or accelerating, but it would have to account for the increase in CR with increased distance.

With SCM light 4B LY away max, due to OCE came 13.4B LY max, from stars now up to 46.5B LY away to date.

With a fixed average rate of CE over the entire span, starlight subjected to CE departing earlier, would take less time to get here than light that departs the same star later. As the later 'flight' encounters more CE on the way, so should have more CR. With STR invariance of the speed of light, without CE, it should take 13B years for light that departed from an object 13B LY away to reach it.[62]

---

[60]en.wikipedia.org/wiki/Visible_spectrum

[61]www.cfa.harvard.edu/~dfabricant/huchra/hubble/ John P. Huchra '08

[62]Einstein's theory https://en.wikibooks.org/wiki/Special_Relativity

Under SCM & SPIRAL CR is evidence of CE at over c. So consider all velocity, duration & sequence of events variables. SPIRAL aligns as well or better, yet SCM makes greater claims:

- Torah Testimony. Bulk of stellar formation very early.
- No 'stretching' of the constellations observations over thousands of years indicates no OCE.
- The Anthropic Principle indicates ID.
- The Universe being Finite and having a Start.
- SPIRAL predicts CR as SPI. SCM reacts to CR.
- If absent, missing 95% Dark Energy/Matter falsify SCM.
- Newton's Laws of Motion: SSSO vs SCM's accelerating CE.
- Radiometric dating without the DTD assumptions.
- Olbers' Paradox regarding a dark night sky and time.
- Our having an (the) optimal View of the Universe.
- The Second Law of Thermodynamics – Entropy.
- Supernovae detectable. Edmond W. Holroyd PhD
- The Lunar Recession Rate. See Danny R. Faulkner PhD
- Earth's Magnetic Field decay rate. icr.org/recent-universe
- CMB. Cosmic Inflation. GRB's. Collimated CR light.
- Lager Structures than SCM predicts.
- More distant objects appear smaller, all else being =.
- No flatness 'miracle' 100+/-x larger the visible sphere.
- The entire universe approximates the visible universe.
- The Earth/Sun Ecliptic being by the center.[63]
- Predicts CR. Predicts CR increase with distance.
- Predicts no CR from any star under 'Y' LY away.
- 1/8th the CE over c as 4B vs 46.5-13.5 = 33B if SCM
- 1504x larger visible universe area claimed by SCM
- 6 days:13.8B Yrs=839B x faster to Adam. Parsimony!
- Hyper-dense Galaxies by CI solves rotation problem.

[63]Ashok K. Singal 'Is there a violation of the Copernican principle in radio sky? 'https://arxiv.org/abs/1305.4134v2 Nov. 15, '14.
Caldwell, R.R., & N. A. Maksimova "Spectral Distortion in a Radially Inhomogeneous Cosmology," arXiv:1309.4454v2 Oct.15'13.
Geocentrism 101 Robert Sungenis PhD Oct.2015.
and 'SPIRAL on CMB' chapters.

See our 'CREST' and 'Blue-shift offset' hypotheses. Either the increase in CR proves we are by the center of the universe, or CR is a function of CE rate and time/distance exposure to CE. If not we find SCM doesn't align with CR increasing with distance, but SPIRAL might still align.

With SCM the distant stars allegedly formed after inflation have been receding from us due to Ongoing Cosmic Expansion (OCE) at above the speed of light for billions of years. If valid, it means the stars pass some of the light they emitted in the direction they are moving.

With SPIRAL cosmological redshift hypothesis this recession of stars at a speed greater than light speed only occurred for a brief period which was during a cosmic inflation expansion event early in the history of the universe.

Metric Expansion hypothesis tries to reconcile a visible universe where distance between objects increased at over c, with the metric, not the object, expands at over c.[64]

So either matter can or did move at a rate above light speed, or CE expanded space at greater than light speed(c), or both. Either way SPIRAL and SCM limit light speed to c.

If one were to assume accelerating OCE is the only explanation of Cosmological Redshift (CR), then SCM might be the only consistent hypothesis. However, SPIRAL is an alternate explanation of CR with no need for OCE.

SPIRAL aligns an initial singularity, proto-stellar formation prior to a cosmic expansion inflation event (SPI), into a stable steady state on the initial oscillation (SSSO).

One may prefer the label big-bang into static finite universe. Not pure static as assume 1 LY a year expansion of CMB and outer edge starlight 'leakage'. Either way the CR aligns with a Earth/Sun ecliptic centric spherical universe.

---

[64] Tamara M. Davis, Expanding Confusion:.cosmological horizons and the superluminal expansion..arXiv:astro-ph/0310808

If valid SPIRAL falsies these hypotheses: An eternal universe, a flat universe, a universe without a center, ongoing cosmic expansion, the notion we do not have the optimal view, and all deep-time dependent assumptions.

SPIRAL predicts CR and where to look for affects of CE. SCM reacts to CR. We provide some calculations for the probability of SPIRAL vs SCM explanations of the cause of the CR and why SPIRAL makes the far more reasonable claims. The far greater claims by SCM requiring a far greater burden of proof. The age of the universe is capped at Years = LY distance to the nearest departure point of visible light effected by CE, including any CR light.

SPIRAL builds on existing scientific law and theory. We are grateful to those who paved the way with hard work and sound thought. SPIRALL CR hypothesis & cosmology model is original to the best of my knowledge.

SPIRAL is consistent with big bang minuscule initial singularity, cosmic expansion, (not OCE), visible starlight from stars now billions of Light Years (LY) away.. 'Classic big bang theory is not about the bang but the aftermath'.[65]

SPIRAL is consistent with Science, ID and YeC. Time dilation is not relied on to reconcile the natural observations with a recent complex creation. SPIRALL is an alternate explanation, to Ongoing Cosmic Expansion (OCE), of CR.[66]

SPIRAL holds creation of an initial singularity, hyper-dense proto-galactic/stellar formation, CI-CE, into a stable steady state oscillation, so no OCE post CI, aligns fully with Scriptural testimony. The 'ShaKai' name of Hashem, alludes to an end of the cosmic expansion, when Hashem said enough! (ShAmar Di - Shin, Daled,Yud)[67]

---

[65]'A golden age of cosmology' Alan Guth (those w/ the gold make the rules).
[66]http://w.astro.berkeley.edu/~louis/astro228/redshift.html (note DTD bias)
[67]Psalms 104:2, Isaiah 40:22..+ pp 76 - Inflated into a static universe.
   Pirkei D'Rebbi Eliezer chapter 3 and Talmud Chagiga 12a Reish Lakish

'Einstein's Doubt', was symptomatic of his biggest blunder: discounting of Torah testimony, so reliance on the deep-time dependent consensus. Einstein's 'Biggest Blunder' a case of premature capitulation, symptomatic of his biggest regret, lack of Talmud Torah.

Einstein thought he was seeing a universe in equilibrium, thus 'Einstein static'. Einstein's doubt stemmed from his failure to reconcile a hyper-dense start, followed by cosmic expansion (CE), to get to that mature density equilibrium, with the prevalent cosmological redshift (CR) of distant starlight. If no doubt, no premature capitulation.

If he considered the solution that is SPIRAL's 'SPI' hyper-dense proto galactic formation prior to CE, attaining mature size and density relatively early, he must have ruled it out due to deep-time dependent assumptions.

If SPIRAL one predicts a minimum of one LY per yr., subsequent to the end of CE, to the nearest departure point, of any direct light at full 'c' standard light speed, arriving at any observation point, that had ever been subjected to any CE. Define this LY radius distance as 'I'.

Now CR, that by definition was subjected to CE, is already readily detectable from stellar objects within 5M LY. So if SPIRAL 'I' = 5M LY max., describes the actuality,

SPIRAL & LCDM competing hypotheses givens:
A light speed limit of 'c' standard light speed.
Most stellar formation had to be relatively early in history.
Visible light departed from a max 400M to 4B LY distance.
Visible CR is due to past CE. If any ongoing CE is disputed.

The max departure point of visible light is 4B LY. If SPIRAL if/when the universe becomes 4B years +5 days old there will be no more CR period. If SPIRAL all age assignments over 5M+5days are falsified, assuming 'I' the nearest departure point of visible CR is 5M LY or lower.

So Einstein's 'biggest blunder', premature capitulation to Hubble's ongoing CE explanation for CR, was caused by assigning 5M+ years subsequent to the big bang. Which precludes consideration of the far more parsimonious & higher probability 'Einstein Static' SPIRAL hypothesis.

Torah testimony, attests to CE ends by end of 4/365(5781) fraction of history, 'Drawn like a curtain' (cosmic expansion) into a 'static tent' At which point, the designer, creator aka G-d of Abraham 'said enough'.

So Einstein's 'Biggest Blunder' premature capitulation to Hubble, was symptomatic of his biggest blunder: reliance on deep-time dependent assumptions, that was symptomatic of what was 'Einstein's final regret' lack of Torah study: that when assigned proper weight, could've avoided the blunder.

Torah testimony is as, or more, reliable than a law of science. Hashem alone had & has 100% perspective. Torah alone stands the test of time, to the extent it covers. Which would have strengthened the resolve of, and guided, Einstein to contemplate SPIRAL, to advance the science far beyond the current deep-time dependent consensus box.

SPIRAL falsifies, for all practical intents & purposes, all deep-time dependent hypotheses & assumptions. So capitulating to the deep-time dependent Hubble's Law: CR is due to OCE, was symptomatic of Einstein's biggest blunder.

From the entrenched deep-time dependent consensus reinforced box, one can't consider SPIRAL, and one doesn't even know that they don't know, the strongest science.

So three elements should have helped Einstein greatly advance the science: Torah study, giving full credence to Torah testimony, and what Einstein would have excelled at: deep consideration how to apply that Torah insight to the empirical observations and pure science. At the pinnacle of Torah & Science alignment is awe, respect and peace.

Based on the empirical observations, if Hubble's Law with ongoing CE conclude over 13B years subsequent to the end of cosmic inflation expansion. If SPIRAL conclude 5M years maximum. So if one assumes a given the universe is well over 5M years old, SPIRAL need not apply. If the universe is under 10B years old Hubble need not apply.

If SPIRAL none of these are true as 'T' is 5M LY max:

- NTD evolution requires over 300M yrs. of continuous life sustaining conditions on Earth. Needs 'T' 300M +.
- Modern cosmology's Copernican Principle premise is deep time dependent, far exceeding 5M cap actuality.
- 'Modern science' is deep-time assumption calibrated. Cumulative, thus faith based. No one's proven anything over 6k yrs old, never mind over 5M or 10B.

SPIRAL not LCDM adds up. No 95% dark energy & matter fudge factor required. Predicts, rather than reacts to, vastly more empirical observations,.. Deep-time dependent LCDM has to fight uphill against the law of science known as entropy.

Thus the far more parsimonious, and far higher probability description of the cosmological empirical observation, that is SPIRAL, should replace the compering hypothesis that is SCM-LCDM, to become the new standard cosmological model. 'Einstein's doubt' references and additional reading:[68]

---

[68] 'Einstein's final regret' Link via Miriam K. @ Einstein archives: R' Aaron Parry quoted in https://jewishlink.news/features/16802-studying-talmud-the-good-the-not-so-good-and-how-to-make-talmud-more-accessible-2
Einstein's 'Doubt' title inspired by Dr. Stephen C. Meyer's work 'Darwin's Doubt'. http://inters.org/einstein-lemaitre via Jonah Lissner PhD
The Universe is not expanding at all. Eric Lerner PhD, via James G. PhD:
www.worldscientific.com/doi/abs/10.1142/S0218271814500588

The 'R' in SPIRAL stands for Redshift (Cosmological Redshift). It is hoped after due consideration SPIRAL will continue to survive scientific rigor, and be accepted as good theory. To realize the vast amount of empirical evidence that is distant starlight attests to the universe being 'thousands.. not billion's' of years old.[69]

The 'A' in SPIRAL is for Attests.
Attests to the SPI = Hyper-dense Proto-Galactic Formation.

'thousands' instead of the precise 5,773 years as of hypothesis formulation in 5773 anno-mundi.,[70] for the same reason Moses told Pharaoh 'about' midnight,[71] even though it would be 'at' midnight.

While the current science caps at the number of years = to the LY distance to the nearest light departure distance of light reaching here now. So safe to say capped at about a million years, we will explain later why 6k rounded is w/in the science sweet spot and expert testing may fine tune it to the year known from Torah testimony.

Rashi points out over confident experts, restricted by the limits of science, would erroneously claim Moses was off the exact mark. The Egyptian scientists did not realize, or did not disclose, their time keeping was not yet exact. How many times has science thought we were close to knowing it all, just to discover more to learn with our next step?

SPIRAL is bolstered by pervasive empirical evidence. With CR light we derive a maximum age of the universe subsequent to a cosmic inflation event.

---

[69]Don De Young PHD, with the RATE team, Thousands...Not Billions, Challenging an Icon of Evolution, Questioning the Age of the Earth. 2005 More on how radiometric dating fails to falsify YeC - RCCF.

[70]R' Zechariah Fendel, Legacy of Sinai , Unbroken History of Torah Transmission with World Backgrounds..1981

[71]Exodus 11:4 see Rashi. Exodus 12:29 'at' midnight

Just like radiometric carbon dating caps the age of artifacts with testable levels at about 70k, soft tissue about 10k, as by then it should mineralized, CR caps the years elapsed to the LY distance to the nearest departure point of CR, less after blue-shift offset and past or present resistance to increased the frequency so lowered the wavelength, and or any past gravitational override of CE, and or the frequency spectrum between center yellow and where the spectrum turns visibly red.

So regular light subjected to a small amount of cosmic expansion, which would be overall the closer distant starlight, at least 6k rounded LY up to a million +/-, during the cosmic inflation expansion 6k rounded years ago, might still be yellow, between center yellow and the start of red. Thus the nearest departure point of light arriving here now, that was subjected to that comic inflation expansion, might be 5779 LY as of 5779 anno-mundi, and not 1M +/- where the nearest CR is already obvious.

See the 'SPIRAL vs Hubble' section 'Exhibit CE' how cosmic expansion and change from hyper to mature-density, would start out slow, then increase velocity with distance.

Assume the size doubled at constant intervals, correlated to the change from hyper-dense to mature density, during the cosmic inflation expansion epoch.

SPIRAL **'Blitz'** on the Cosmic Distance Ladder.
Illustration of a fourfold change in density.
Parallax of stellar objects subjugated to cosmic expansion, reflect
current, not their light departure point, distance.

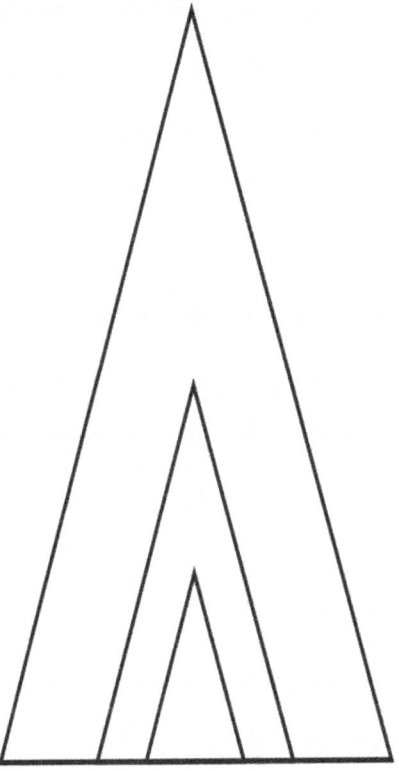

Baseline parallax 0.5, 1.0 & mature density 2.0 AU. Not to
scale.1.0 LY = 63,241 AU. Reflecting ONE stellar object shown
when at I, 2I & 4I LY. During the cosmic expansion epoch. That if
SPIRAL ended day four. We observe the light by the baseline at
current universe density.

So parallax is constant for a stellar object now at 24k LY,
as when 12kLY, & 6kLY distance from us. If 5,782 years elapsed,
the light departed 5,782 LY away & shifted due to the metric
expansion of space, so measure to current 24k LY.

With SPIRAL, hyper-dense proto galactic formation was PRIOR to (near all, based on volume) hyper cosmic expansion (CI). The density factor during CI can fully align with a 5,781 year lapse, and max light year (LY) travel distance, to date (as of Sept. 17th, 2020), of light, (& CMB), visible here & now.

With all of the visible universe beyond Radius 'I', the border where (after adjusting for orbital peculiarities), on this side we see stellar objects 'normal light' (that was never subjected to any cosmic expansion), and beyond which all stellar objects, visible here and now, light was subjected to at least some cosmic expansion, during CI.

That CI began with the universe hyper-dense and ended by the universe at mature size and density, by the end of $4/365(5781)$ a fraction of history.

Now with the mature density at up to 4B LY. The maximum light departure point of the most distant stellar objects, be it SPIRAL, with no ongoing cosmic expansion, or SCM-LCDM with $46.5 - 4 = 42.5$ B LY of subsequent cosmic expansion over the course of an assumed 13.x year.

99.99% of viable stellar objects are beyond 'I' and we see all from when they were at 'I'.

As anything light departing from beyond 'I' has not had time to reach us yet. Due the light speed limit of standard light speed 'c'.

Thus a 'blitz' of all visible light here and now, by all stellar objects beyond 'I', from 'line of scrimmage' radius 'I'!

Now the more distant the resting distance by the end of CI 'day four', the higher the recession velocity, denser, and the earlier in the day, the light we see here and now departed the object, as it was at distance 'I'.

All else being = the more distant a stellar object the dimmer, it appears. Density/intensity & rate of recession are variables if we see two like objects from distance 'I'.

Any 'new' visible stellar objects, that began emitting light after day 4, have to be w/in 5781 LY. As their light couldn't have reached from beyond 'I'. Any 'new' ones are w/in 5781 LY. Others were emitting light within 4/365(5781) a fraction of history.

Accurate LY distance estimates may depend on the model and how well the assumptions match the actuality. A straight line being the shortest distance between two points.

So the more bending the longer it traveled and takes to get from it's departure point, which may not be where the stellar object ended up by the end of CI.

If the angle to a point 1k LY a hairbreadth, 2k LY far less. So considering the max. light could bend and uncertain departure point of the light, under 1k LY is the parallax distance we have more confidence in. Past that we lack true depth perception.

A massive discrepancy in outcome depends on a minute angle. A shredded thread isn't enough to hang a hat on, never mind an alleged universe 93B LY across the visible part alone.

Most of that hinges on the assumption of ongoing CE. Which depends on the assumption of Dark Energy. If SPIRAL the entire universe aligns with lights 4BLY max departure point.

How standard are standard candles? 'Despite a consensus the north star is a 'mere' 434LY away. It may be 30% closer which, if correct, is especially notable because Polaris is the closest Cepheid variable to Earth, so its physical parameters are of critical importance to the whole astronomical distance scale.'[72]

[72]http://en.wikipedia.org/wiki/Polaris
    https://en.wikipedia.org/wiki/Cepheid_variable

Cepheid Variables: The greater their 'resting distance' beyond radius 'I', the denser and faster it was receding, where the light we see from it here and now, departed at Radius 'I'. So the pulsation 'period' may not correlate to the absolute magnitude of it's luminosity (brightness).

If receding at 1 light hour per second during Hyper Cosmic Expansion, that reduces it's 'apparent' from our vantage point, pulsation frequency, and increases the duration of each pulse, by a factor of 60(60)=3,600. Yet does nothing to increase or decrease it's absolute luminosity, pulsation frequency & duration. So to understand the 'period' pulsation frequency, duration, luminosity, .., calibrate for rate of recession and density at 'I'.[73]

Gamma Ray Burst (GRB) from supernova are assumed to be uniform, then employed as standard candles. This may also prove inexact. If snowflakes are rarely identical, how much more so stars?! Even the cause of GRB's may be revised if SPIRAL.

The 'Inner Universe' (IU) our sphere with a radius 'I' to the nearest light ever subjected to any cosmic expansion departure point. The start of the 'Outer Universe' (OU) .

Due to CB.., the nearest object with CR light is nearer than where the CR becomes pervasive, but further than the 5,781 LY distance departure point. SPIRAL predicts the IU radius grows one LY per year. With no ongoing CE, this LY radius of normal light is equal to years elapsed since the end of CE/CI.

If our 5,781 LY radius prediction based on RCCF, to where the inner and outer universe meet is correct 5,781 LY out of 13.8B LY distance deep-time doctrine estimates would predict based on 1LY per year, is very significant statistically, thus if and when verified, attest to SPIRAL.

---

[73]https://lco.global/spacebook/distance/cepheid-variable-stars-supernovae-and-distance-measurement/

Even if we assume certain supernova, are exact standard candles.[74]

We'd need to observe them at the critical timing, distance, and intervals, .., to measure to the nearest, with light ever subjected to cosmic expansion.[75]

SPIRALL is testable. Under SPIRAL one would predict no subjugation to cosmic expansion from any visible light with a departure point of under 5781 LY minimum radius 'I' to date. Then a negligible amount of subjugation starts at radius 'I'. As the subjugation to CE is cumulative. The cumulative subjugation at 'I' +1B LY is 1k that at 1M LY is 1k that at 1k LY. So an overall increase with distance, with 1M fold CE subjugation at 1B LY compared to 6781 LY.

Visible CR set's in about where to expect if SPIRAL, after calibration for CB offset and spectrum prior to Red.

The current cosmic distance ladder (CDL), may overstate distances. Consider max reliable parallax and no OCE. That distance also puts a cap on years elapsed post CI. Both SCM and SPIRAL hold a CI epoch relatively soon after the start.

Macs J1149. A Gravitational lensing factor of 10 alleged. Could be mostly CB offset at play? [76]

Narratives that require more to get to less but do not explain where the more came from, cause greater unexplained gaps, take more energy, are less probable, require greater faith, then the RCCF – YeC.

For example, due to the law of entropy, it cannot be we can repeat a cycle of singularity, to expansion, to singularity, to expansion, without losing something.

So does claiming extraterrestrials, as that increases what we do not know, such as how they came to be, and how they came to the point they could get us to this point.

---

[74]Measuring..with Supernovae, Saul Perlmutter, arXiv:astro-ph/0303428 03/'03
[75] Edward Hubble "A relation between distance & radial velocity among extra-galactic nebulae". Proceedings of the NAS 1929
[76]www.nasa.gov/feature/goddard/2018/hubble-uncovers-the-farthest-star-ever-seen

Multi-verse narrative only adds to one's dilemma. A pure scientist who proposes this type of hypothesis is telling us they have yet to consider RCCF framework and SPIRAL.

To assert our universe is the product of an even greater physical one, just as a small rock can come from a larger rock, is kicking the can down the road away from an answer, as an even greater universe is a greater claim.

The greater the claim, the greater the burden of proof. Which reminds me of the anecdote of an easy way to make a small fortune - start with a large fortune. So now they cannot credibly explain how either our universe, or the other alleged one/s, came about.

SPIRAL is consistent with both cosmic inflation expansion at above light speed (c) with light itself not exceeding the speed of light. Despite cosmic expansion, light itself can't exceed the speed of light. Perhaps due to light not being subject to inertia. That means if the matter that moved away/apart at speeds above c, due to CE, or anything else the light speed is limited to 'c'.

Light emitted from a stellar object receding from the observer due to cosmic expansion, will accrue CR with the distance subjugated to the CE.

Such as the 'day 4' Hyper CE-CI of SPIRAL, by the end of which the universe attained mature size and density, so gravitational bound equilibrium, of the entire universe.

Factor in the extent and distance, if any, the stellar object was subjugated to SPIRAL's galactic near side 'cosmological blue-shift offset' during CE.

The Cosmic Distance Ladder & Parallax triangulation. Calibrate for: Change in density if SPIRAL and for gravitational pull by the sun on the light. Angles that extrapolate to a point 5.78k to 30k Light Years (LY) away can in actuality be from light that originated from a distance of 5778 LY or closer. 'maximum reliable parallax is a few hundred LY'[77]

See 'SPIRAL vs Hubble' Exhibit CE how/why proto hyper-density was a far greater factor based on cosmic expansion.

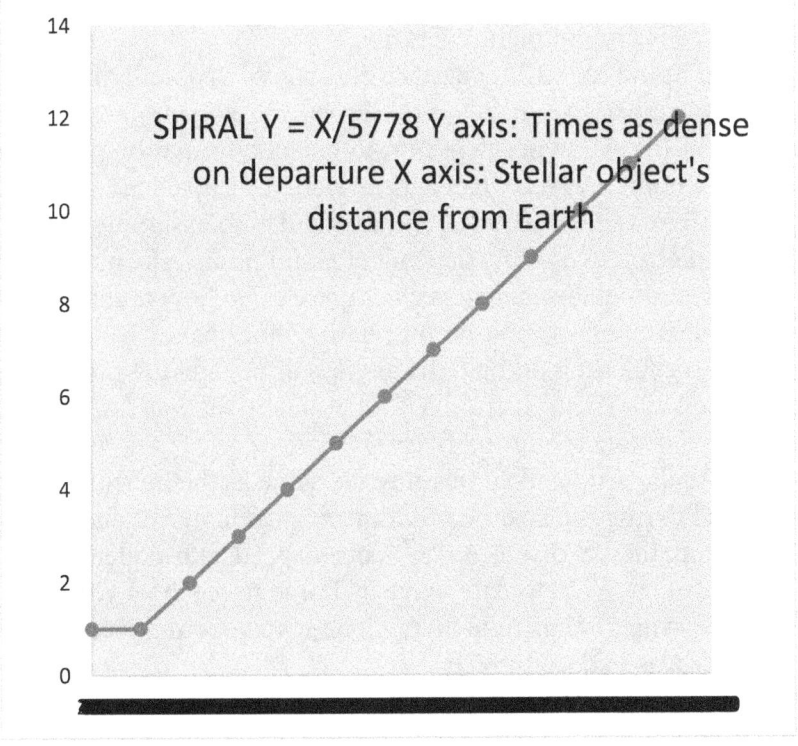

SPIRAL Y = X/5778 Y axis: Times as dense on departure X axis: Stellar object's distance from Earth

Chart Title = Times stellar objects were denser at their visible light's departure. The increase starts with stellar objects beyond 'I', as of 2018+3760 was 5,778LY. Y = X/5778.

This chart shows up to 70k LY.

[77]'Focus on physical science textbook' Jay M. Pasachoff PhD

Y = Times denser on visible light departure than at stellar objects mature density at end of Hyper Cosmic Expansion (CI) epoch.

X = LY distance stellar object was subject to Cosmic Expansion up to about 2B LY if SPIRAL.

If SPIRAL change in density during CE masks departure distance of light from objects ending up over 5778 LY.

If SPIRAL change in density during CE caps departure distance of light from objects over, to, 5778 LY to date. X = LY up to 5778 is current mature density.

If Spiral max LY departure distance of visible light here/now capped to amt. = to yrs elapsed subsequent to end of CI.

If SPIRAL hyper-dense proto-galactic formation pre-CE a departure point 5.78k LY can appear were the object ended up, be it up to 30k based on parallax and extrapolation assuming no margin of error, deep-time dependent assumptions, the most precise instrumentation, and perfect guesses on how much gravity bent the light, and calibration for relative motions..

So even after all that, the angle will be indistinguishable if it departed 30k LY, or if from 5778 LY, due to the increase in density w/ distance, starting over 5778LY.

As the ratio and angles stay the same as the density increased during the cosmic inflation expansion event, and only light within the LY distance = to years elapsed, subsequent to that event we put at 5778 to date, have had time to reach us yet at mature densities. Thus light arriving here now from under 5778 LY, never was exposed to CE.

'..gravitational lensing and the amount of bending is one of the predictions of Albert Einstein's general theory of relativity. Classical physics also predicts bending of light, but only half that of general relativity's...In general relativity, light follows the curvature of spacetime'[78]

Let us use as a visual aide a fountain of water.[79]

- The water is also subjected to gravity.
- Consider the most distant visible light source subject to parallax as the zenith at the top center of the water.
- Now consider water from somewhat closer to the ground also comes in at the same or closer to the same angle by the time it gets to the ground.
- Now consider the sun as being in the center or the water where it hits the ground.
- Consider the edge around the base of the water where it hits the ground as the Earth ecliptic around the sun.
- Now consider we see the light / water only when it gets to the ground, not it's departure point.
- If not calibrated for gravity the angles we see/use for parallax/ triangulation would result in a water high point much higher than the actuality.

Now with starlight the assumed departure start point of the light will affect how much we calibrate for gravity.

Not only the assumed initial trajectory and start distance but other assumptions like other mass, so other gravitational force that might mitigate or increase bending, along the way. If the starlight is heading straight down toward the sun the gravity shouldn't bend the light at all as it is pulling it directly toward it.

---

[78] https://en.wikipedia.org/wiki/Gravitational_lens
[79] wikimedia.org/wiki/File:Fountain_at_Linn_Park_Birmingham_AL

Then if a small slant there is not much it can bend it. The greater the initial angle the more possible effect. Therefore, to assume the light is coming from 30k and to calibrate for gravity on an angle of 30k is circular reasoning. So, e result doesn't prove anything, other than it is one possible result that may not hold up. The actuality of how far away the light departed could be 5,778 LY or much less.

The 'hairbreadth' fraction the angle for triangulation to 5,778 LY is so minuscule it would be easy to understate the effect of gravity and fail to catch the error overstating the distance. Less than a hairbreadth the angle if 30k LY. A shredded thread isn't something we should hang our hats on.

The base line being 2 AU, an AU being about 93 million miles, that is the average distance Earth to sun.

- An object 'just' 1LY away is 63,241AU (a tiny fraction).
- An object 1000 LY distance is 63 million rounded AU.
- An object 5777 LY distance is 365 million rounded AU.
- An object 30,000 LY distance is 1.9 billion rounded AU.

5280 feet in a mile, so substitute feet for AU so 1 foot is to 12 miles as 1 AU is to 1 LY.. (rounded), to help envision the angle.

1 foot is to 12,000 miles as 1 AU is to 1000 LY.
1 foot is to 69,324 miles as 1 AU is to 5,777 LY.
3 inches is to 8,666 miles as 2 AU is to 5,777 LY.

So don't hang any hat on parallax measurements being absolutely exact starting at _x_ LY or so, and less reliability with distance, all else being =.

Assumptions on where the departure point of light as the departure point /trajectory is a variable on if/how the sun ..will bend the light?

How do we know which light was bent and how much without assuming the start point where it starts to have an impact?

How much can our gravitational field attract bend the light at all. does our magnetic field deflect the light?

Consider the northern lights phenomena, space-time curvature, refraction, relativistic beaming, axial precession, aberration of light and gravitational lensing.

Standing still close one eye, than the next, at what distance do you no longer notice a change in angle between an object and it's backdrop, using our eyes 3 inch +/- baseline parallax? Without context/reference points, is it where we lose depth perception?

Try driving x MPH while spinning. Compare this to the 2 AU reference frame and distances under Parallax as Earth is moving 1M MPH and rotating xMPH when measuring. Where we only view a tiny fraction of the tail end of the journey of the light. So, no direct observation of most of the trail trajectory, distance and departure point while factoring gravitational bending, motions, reflection...

A human observer seeing the sun and moon for the first time w/zero context of ionosphere lensing .. and known reference points could only speculate on the exact size and distance. How accurate is our best survey device to what distance with just 2x12 inches of baseline parallax? Our most precise devices have limits.

By a solar eclipse we see the moon is closer and smaller as they appear the same size. It took a while till we determined the sun averages about 93M (1 AU) miles distance from us, about 400 times rounded the distance (and size) of the moon, whose 232k+/- distance fluctuates by over 30k miles.

So as the crow flies, we go 2AU every 6 months 2(93M) /180 days = just over 1Million miles a day. As we circle the sun the average speed is 67k MPH = 1.6M a day. We also rotate depending on our distance to Earth's axis at up to 465 MPH. We also go 140+/-? MPH with the sun along the Earth-Sun ecliptic.

If so accurate, why large disparity estimates to even relatively close stars? Stationary north star Polaris 99-133 parsecs, wanderer 'Betelgeuse' 222+48/-34 margin...

If the relative speed and motion of the Earth and the Sun affect the trajectory of the light, is that calibrated in? How do we know with certainty how large and bright the distant visible stars/galaxies are? All assumptions and variables to deduce distance this should be disclosed.

If one assumes deep-time dependent SCM, one assumes OCE and non-existent missing dark energy and matter in the mix/mass. Results are valid only if the underlying assumptions and building blocks are correct.

Don't assume uniformitarian assumptions. Do you assume a stellar object 15k LY, 1.5M LY, 1.5B LY.. has always been at that approximate distance?

Obviously if it was subjected to a material amount of cosmic expansion it was not. So, the light we see here now from all objects at and over 5778 LY, could have departed when 5778 LY distance.

If SPIRAL it is safe to assume, only objects under 5778 LY distant to date, the light arriving here now, wasn't subjugated to CE and departed the object when the mature post-cosmic inflation expansion density.

Did you factor for change in density? If SPIRAL, stellar objects now 11,556 LY distant were 2x as dense and 50% closer when their light, visible here & now departed them. Those 23,112 LY distant were 4 times as dense and 1/4th of that distance, when their light visible here & now departed them.

Think of the leavening raisin dough analogy, but where the raisins themselves are expanding. If the proto-Earth and proto-sun also expanded in proportion to the stellar objects & space, the parallax angles remain constant, and an object now 23,112 LY distant based on extrapolation of a straight lines from the base to the object, would come in at the same angles as when it was 5778 LY distant and ¼ the density..

Think of a 30 foot fountain where the water jets depart the ground and there has only been time for the water to reach the summit, but we are unaware of that. That our view is limited to under 1 inch off the ground.

So too our view is limited to light as it reaches here, and if SPIRAL no light that departed over 5778 LY to date has had time to reach us. Despite light from the most distant visible objects, even up to billions of LY, does reach us, as it departed when they were closer.

In conclusion disclose assumptions in using parallax to build a cosmic distance ladder. Are you factoring in for the change in density on objects over 5778 LY to date, if SPIRAL? Also consider how and where light may have been bent by gravity..

Sure bending means the distance light would have to travel and time it would take that light to reach us is greater than if a stationary objects light departs and reaches us in a straight line.

Yet we lack the context/perspective to know if the light departed from where the object is now, or if from where it was during the cosmic inflation expansion event. If SPIRAL 5786 LY distant to date (2025/2026)

The enormous difference a tiny fraction of an angle makes with even 1LY + increasing with distance.

The actuality may be 5,776 LY is the most distant visible light departure point to date, as of 2015, as SPIRAL predicts, with an increase of 1 LY per year.

See 'The Recent Complex Creation Framework' for understanding science in maximum available context how/why the strongest science aligns with ID & YeC.

See 'Torah Discovery Chronology' (TDC) alignment of scriptural testimony and ancient civilization for the tightest unbroken time-line years to date from creation.

The highest probability explanation of empirical observations we define as the stronger science. SPIRAL is the stronger science.

With Torah testimony the why and when is a given. For what reason deep-time? It fights the laws of science and we're absent for the vast majority of it. No reason! Why design the universe so that we would never see beyond a fraction of it? No reason!

SCM's vastly greater claim 46B LY radius visible universe, a fraction of the entire universe, w/ OCE so Hubble event horizon means we'll never see any light ever emitted from any galaxies beyond a 14B LY radius.

SPIRAL's parsimonious design for optimal view of the entire universe makes sense & stronger science.

In conclusion: After 100 years of increasing estimates, we've reduced and capped the size of the universe. Some hold at the entire, at least 500:1 the visible, universe, many hold at least 200:1, SPIRAL finds near 1:1.

We cap the visible universe radius to the consensus original departure point of the most distant detectable stellar objects, so 4B LY maximum. Expect the actuality is closer to to a 1B LY radius.

Down from the 5M%+ times larger entire universe, 500:1 a visible universe of 46.5B LY Radius. It is still a massive area. Still accounts for 100% of the empirical observations. We also find the '5%' 'normal matter' approximates all matter.

Just using different lens to understand better what we are looking at. The far greater Parsimony and design should make us appreciate it even more. Knowing it's is all for us, to grow and connect w/ our Father and King, helps us appreciate ourselves and our fellow, so leads to peace.

At 200 times the visible universe of SCM 46.5B the entire universe area is 2.5M times greater than if SPIRAL.

Cosmic Distance via Spectrography & Photometry:
see later in SPIRAL on JWTS-JADES and Lyman Break.

Cepheid Variable research help request:
Using Cepheid variable links below please identify what material correlation, if any, between pulsation duration, and/or frequency, and distance. TY, rmp. links via Krister Sundelin:[80]

---

[80]www.astro.utoronto.ca/.../cepheids/cepheids.html
http://ogledb.astrouw.edu.pl/~ogle/CVS/
https://heasarc.gsfc.nasa.gov/W3Browse/all/gcvsegvars.html

The LL of SPIRAL(L) stands for Lagging Light.

So SPI-RALL = Hyper-dense Proto-Galactic (Stellar) Formation Preceded Cosmic Inflation expansion – Cosmological Redshift (CR) Attests to their Lagging Light of those stellar objects that were subjected to distancing from us during Cosmic Inflation.

As the CR is the result of, and what we could predict, based on stellar objects receding from us during cosmic inflation expansion epoch.

The CR is Lagging Light (decreased frequency/longer (increased) wavelength) as it arrives over a longer span than if no distancing.

For example light from a stellar object that recedes xLY will arrive over a span of x years more, than if it hadn't receded.

If the sun which is 8 light minutes distant, has a linear distancing, due to cosmic expansion, receded 1 LY during the next 1 minute, the light it emits over the next 1 minute, will start to arrive in 8 minutes and finish arriving in one year and one minute from now. During which time all the light that arrives from it has been subjected to cosmic expansion, so is should appear with slight but uniform, cosmological redshift. After said span we will start to get normal light from it again.

If ongoing CE, as it does per SCM, it accumulates, with $46.5B-4B = 42.5B$ years and light years' worth of LL accumulated over 13B yrs increasing at about 3.26 annually.

In SPIRAL no ongoing CE, so up to about 2B of LL.

If CR is due to ongoing, not past, CE, it raises issues. SPIRAL vs Hubble and CREST hypotheses examine the increase of CR with distance of the stellar objects.

Under those conditions how did any CR light much over the current 1.69 z (CR redshift) unreachable limit arrive here in the first place?

Why would the future visibility limit of 19,000 MPC (3.262 LY per Parsec times a million) be much higher earlier in history than it is now?[81]

Neither a problem under SPIRAL as no OCE, so distant light heading our way gets here in the number of years equal to the LY distance to go.

SPIRAL helps provide context to advance in science. SPIRAL may be the best description of the age, formation and structure of the universe. An infinite universe doesn't add up. A finite universe and the conservation of energy law attest to the first cause being supernatural. The laws of nature, and time itself, also depend on that supernatural first cause.

The Moshe Emes RCCF framework explains how we know the first cause is The One designer/creator, aka G-d of Abraham, just as described by Moses.

Our motive has zero scientific weight, as a hypothesis adds up to a probability based on it's own merits.

That all matter started in a singularity is one of many indicators from which we can deduce Hashem is One.

Time started when the energy for the initial primordial matter for the 'big bang' came into being, so Hashem predated time and therefore always 'was'.

Per the RCCF framework the physical universe would revert to nothing if not for Hashem who therefore 'is'.

Time will cease to exist if and when the physical universe reverts to nothing. So Hashem also always 'will be' for at least as long as the physical universe and time exist.

All physical extant matter was contained in the initial primordial matter. As King Solomon aka Koheles taught 'nothing new under the sun'.[82]

---

[81] www.astro.princeton.edu/universe/ms.pdf
[82] Talmud Berachot 59a

The primordial matter started in super dense tiny size. RambaN wrote this 800 years ago based on Talmud redacted 1600 YA and recently popularized by / in big bang theory.

If one can create a force of nature from scratch one might retain the capacity to manipulate said force. So why wouldn't Hashem retain such a capacity?!

Gravity, the weakest of the four forces may have never caused it alone. Both SCM & SPIRAL concur the attractive and repulsive forces needed for star formation and for inflation were inherent in that initial singularity. If the four forces - gravity, electromagnetic, strong and weak atomic forces, were not always distinct as they are now, as in 'Grand unification epoch/energy'. Hashem would have manipulated them at will during creation week.

Now as we illustrated prior with parallax staying constant, for a stellar object due to being subjugated to comic expansion, as it is the metric expansion, fabric of space that is causing the distancing. Think of the classic leavening raisin dough analogy, accept with SPIRAL the 'raisins' ie proto-galxies are also expanding.

So stellar objects that began 1, now 24k, LY away, the approximate distance they attained at the end of the hyper-cosmic expansion epoch, when the universe attained mature size and density, kept a constant parallax the entire time.

Hypothetically if we had 100% accurate parallax perspective of the entire visible universe. If SCM the most distant visible stellar objects that departed 4B LY distance, that traveled 13B LY, starting 13B years ago are now 46.5 B LY distance. We see on the baseline here and now, so based on the 46.5B radius. So the parallax should is 46.5B LY not 4B LY light departure point and, not 13B LY. If SPIRAL their light that departed 'I' LY distance, and years ago, gives a parallax of 2B LY.

SPIRAL's **'GRIP'** hypothesis

Vortex like Galactic Rotation Indicates Primordial Electro-Magnetic Radiation (EMR) fields obviates the need for Dark Matter.

SPIRAL's Hyper-dense Proto-Galactic formation prior to & during hyper-cosmic expansion explains galactic rotation data without MOND to solve without the 'Dark Matter' fudge factor.[83]

Gravity pulls on mass and light. Beyond radius 'I' we see from mid-CI, day 4, when at radius 'I', 5781 LY to date distant & yrs ago. Not where they came to 'rest' by the end of that CI day.

Cumulatively stronger than gravity is EMR. In SPIRAL's Magnetic Repulsion hypothesis EMR was the key agent for CI. During the hyper-dense phase the EMR, was likewise hyper-concentrated. Is CMB not the residue of that EMR agent of CI? We detect now from a departure distance 'I'.

EMR and Gravity, two agents that caused the 'vortex' of visible galactic rotations. Expanding the universe from a radius of 1LY rounded up until equilibrium at 4B LY max mature size and density, by the end of day 4. The more distant the galaxy, the more dense, and earlier it passed the radius 'I', during CI day 4.

Think of walking a fixed speed and time around a track. Ten laps of 100 yards per every one lap if 1,000 yards. If so during the CI phase a galaxy 1M x denser might rotate 1M fold.

Even if a constant RPM (revolution per minute) regardless of density, depending on how fast they were receding during CI is a variable on how we perceive it. Even if spinning @60RMP CI at 3.5 light days per second. We see it revolve twice a week.

---

[83]If SPIRAL no need for Mordechai Milgrom's complex Modified Newtonian Dynamics 'MOND'. https://arxiv.org/abs/2210.13472v1 Pavel Kroupa. SPIRAL is the alt. to Dark Matter and MOND Prof. Stacy Mcgaugh seeks! with Dr. Keating 34:00 in at www.youtube.com/watch?v=NnpyFk2WIME

So no deep-time dependence required. Giving us greater insight behind Period Luminosity-Cepheid Variables aka 'Leavitt's law'.

Past SPIRAL radius 'I', overall the more distant a galaxy from our central view, the earlier day 4, the more dense it was where, we see it from. The greater the subjugation to CI. Use size, luminosity and CR.., to estimate 'resting' distance. Where size, distance & speed are also disputed.

By observing the Galactic Rotation now, one might be able to compute the time-lapse and expansion rates that work if SPIRAL. Hyper-dense galactic formation, thousands, not billions of yrs. ago, is compatible with stellar orbits and velocity w/o DM.

We've looked for, but not found DM. The highest probability is it is non-existent. Assumption based illusion till proven otherwise. Like missing transitional fossils predicted & required if NDT valid.[84]

Galactic rotation, velocity & orbits, force SCM galactic & stellar formation, via gravity over deep-time, to require DM.

'If we haven't detected the missing dark matter (DM), 20 years from now I may be concerned'. Why wait 20 (more) years? Didn't someone say that over 20 years ago? :) [85]

As there's a readily available explanation waiting, just outside the consensus box, that can add up and cross check with basic science and all empirical observations – SPIRAL.

---

[84]Paval Kroupa on Mordechai Milgrom 'MOND' The Dark Matter Crisis 'Darwin's Doubt' by Stephen C. Meyer's PhD

[85]'Why do astronomers believe in dark matter?' https://phys.org/news/2019-09-astronomers-dark.html Michael J.I.Brown PhD

These same empirical observations may attest to SPI = Hyper-dense proto-Galactic formation by/before CI. If each hyper-dense galaxy was as an egg in a cluster, each galaxy composed of smaller 'eggs' clustered. The more uniform, closer, start, along with thousands not billions, year time scale, obviates the need for Dark Matter (DM).

'GRIP' also obliviates the need for Dr. Hartnett DM alt.[86]

To compare apples to apples assume: Standard light speed 'c' constant for all. The laws of physics are universal. Could laws of physics differ? Maybe, if any singularity, which need not have been, or be, if SPIRAL's 'Black Hole illusion Resolution'. Merging Neutron stars, movement at near light speed.[87]

Near instantaneous stellar formation is possible.[88]

'Primordial EMR Fields' so called by Brian Keating PhD, shows SPIRAL's 'Magnetic Repulsion' for cosmic expansion, & 'GRIP' hypotheses, are based on a viable premise.

Just as CI started and stopped under SCM, one can hypothesize conditions for CI if SPIRAL to have stopped. Use a microwave popcorn or magnet deflection analogy.

Whereas the presence of all the dark matter required for SCM has not been proven present, and is hypothesized based on the assumption SCM is the actuality. If it is really there does it preclude other consistent hypotheses?

---

[86]Why Dark Matter everywhere - creation.com http://creation.com/why-dark-matter-everywhere#.XMsdqsoKgEQ.twitter John G. Hartnett March 31,'15

[87]'On their own, neutron stars can spin so fast — up to about 2/3 'c' ..create the largest known magnetic fields in the Universe '.. also emitted is matter. Merging Neutron Stars form Unstoppable Jet, moves at near 'c' Siegel 2019

[88]https://arxiv.org/abs/1602.01985 On the inconsistency between cosmic stellar mass density and star formation rate up to z-8H. Yu,F.Y.Wang

Often in physics and chemistry if there is going to be a reaction, it happens, or can happen, right away, and there is no long gestation period requirement.

Chazal: 'luminaries started on day one and placed on day four'. How Ramban hypothesized the initial split?

The reference in Ramban is in 'Genesis and The Big Bang' by Dr. Schroeder. Perhaps Ramban envisioned the split on day two. Think of a single cell as representing all matter, then divides and separates into 2 complementary projects that undergo expansion, as in the analogy of two warps of thread.[89]

Ramban and Rashi hold all matter was created on day one. The sages and Rashi hold the proto-stars were already encoded by the end of day one. I'm not aware of Torah or science that precludes SPIRAL's proto-stellar formation before 'cosmic inflation day' four.

The proportion of time elapsed prior to CI if SCM relative to that if SPIRAL, is reasonable, considering the disputed processes involved if one or the other. Everyone who holds proto-Stellar formation Preceded CI (SPI), should be happy to incorporate the rest of SPIRAL (RAL) into their view.[90]

The Earth / Sun Ecliptic is the approximate center of the spherical visible universe. If SPIRAL is true the approximate center of the whole universe. As the whole universe would approximate the visible universe.

One of the functions of the sun is to act as a central heating system for Earth. The molten core of the Earth might also be considered a central heating system. So when we say we are the center of the universe, it does not mean the absolute physical center, just as we are not in the center of the Earth, or fixed in the center of our solar system.

---

[89]Talmud Chagigah 12a
[90]Genesis 1:14 see Rashi.

Like at home our use might revolve around the living room, the library, the bedroom, the kids' room, the kitchen or the bathroom, not the physical center.

As another analogy, think of the Mediterranean Sea as our ecliptic, a cruise ship as our Earth, the ports of call as the 12 main constellations. We passengers visit all the ports, just as the Earth visits the constellations all around the ecliptic, annually. Daily if we could see through the daylight.

The sweet spot of a baseball bat isn't its center. Earth is at the sweet spot of the physical universe for mankind.

Favoring galactic formation, that was early in the universe, being 'thousands not billions' of years ago 'Star Clusters Fall Apart Faster Than Expected.'[91]

SPIRAL explains why the observational science of CR places us near center of the actual universe, that approximates our spherical observable universe.

If in fact we occupy a privileged location in space and time AT THE CENTER of a spherical bulge of matter and curvature, then it may be possible to explain a vast catalog of observational data without the need to invoke new physical effects such as dark energy. See SPIRAL on CMB chapter.[92]

---

[91] Sabine Hossenfelder is Dark Matter real? 5:52 https://youtu.be/U4sw3-__pGo
[92] Caldwell, R.R. & N. A. Maksimova, "Spectral Distortion in a Radially Inhomogeneous Cosmology" arXiv:1309.4454v2 15 Oct '13

If the Earth is not the actual approximate center from which the Pearlman SPIRAL transpired round about:

- An optimal view of the stars from our solar system and the vertical alignment of major cosmological features to the ecliptic 'would be highly improbable.'[93]
- Why did light allegedly start out 13+/- billion, years ago and light years distance, not bend, scatter, dissipate and or get blocked, more?
- Why did a universe that started at & expanded from a tiny point, not have that point nearby its center?
- Why is the CR consistent in every direction if we are not by that center point?
- Why are there approximately the same number of stars, within xk LY orbiting toward, as away, from us?
- If SCM the solar system should have been torn apart by 'The Big Rip'[94]. or 'crunched' long ago.
- How did distant galaxies and any super massive black-holes, form so early in time if not SPIRAL?
- Why are the massive structures detected at the far ends of the observable universe larger than had been predicted under SCM if not SPIRAL?
- Why are the asteroid belts so well positioned around us, and missing so much matter? Is it perhaps from where the other celestial bodies emanated?
- If SCM with OCE why is light from stars 46B LY that departed 4B LY away 13B YA not dissipated?
- Despite dynamics, the sun has long kept a stable size and distance of around 400x that of our moon.
- Note size/distance relationship, and stability of both.

[93]Daniel Shafer PhD arXiv:1312.1688v2 2014 + 2007
   www.personal.umich.edu/~huterer/PRESS/CMB_Huterer.pdf
[94]Caldwell, R, Kamionkowski, M, Weinberg, N. "Phantom Energy and Cosmic Doomsday". arXiv:astro-ph/0302506 2003

If true, SPIRAL falsifies all deep-time dependent scientific hypotheses, like Neo-Darwinism (NDT) and SCM.

Keep in mind just because one narrative is true, and the other is not, does not mean some bits and pieces in the falsified narrative are not true.

For example, just because adaptation and speciation are true, does not mean NDT is. As ID & YeC models may also include same.

RCCF framework of six principles for understanding science in maximum available context is an internally consistent Intelligent Design (ID) & Young Earth Creationist (YEC) model alt. to deep-time dependent doctrine models.

An ID advocate still holds ID with SPIRAL.

A YeC should gladly embrace SPIRAL to the extent it merits, as it independently falsifies all Deep-Time narratives.

An engineer should appreciate SPIRAL over SCM as from a design perspective the less time, energy and or space required, all else being equal, with the same result, makes more sense. 'Parsimony' & 'Occam's razor' come to mind.[95]

If you had a choice would you build a super-computer of = durability/functionality the size of a building or lap-top? Over a day or a decade? using a billion times more material than needed?

So would not G-d create this universe by SPIRAL 800B times quicker! in 1504 times less area! So no requirement for missing dark energy/matter so 1/20th the material! So no wasted flattening of the universe with 100x+/- beyond use! Providing us the optimal view as we can see 99% more of the universe than from any other distant viewpoint! While providing overwhelming empirical evidence that is distant starlight trails of the validity of Torah revelation testimony and timeline! So this work is signed and dated. Even if G-d could, there is zero reason to do by SCM, just to provide the exact same natural observations.

---

[95] T' Shabbos R' Yehudah quoting Rav, no needless creation. Purpose based.

Perspective is important. For example, with context we know the tires on each car weren't manufactured in each individual parking place a car is found. With more context we find they didn't originate at the tire store either. The manufacturer gathers and intensely processes the base components. From there the tires are distributed widely.

So just because a Galaxy has a lot of stars post inflation, that may not be where the proto planetary-disks / proto-galaxies / proto-stars formed. After all, galaxies appear all around the universe, and it is a lot easier to centrally manufacture them and then distribute them. As it is more complex, and takes more energy and matter, to produce and place just one beneficial star, than all the tires in the world.

If you hold by the SCM you should hold all the matter and information required to form all the stars, came from the same place, which was the initial singularity. A material greater amount of time elapsed and energy was required, to go from that same point, to what we have now, than if under SPIRAL. So SCM makes the greater claim, thus a greater burden of proof on SCM.

If you hold by the big bang theory with its common origins of all existing physical matter being from the initial singularity of primordial matter, the SPIRAL is reasonable.

It reminds me of the dispute as to whether all life evolved from the first living organism by chance, as claimed by NDT, or did One common creator make a couple of hundred base Min/kind types in full stature, with the designed in ability to adapt. Taken in context, the latter is a vastly greater probability, than NDT, if NDT where even possible at all.

It is not legit to claim the mantle of science to assume all the required conditions to support a 'scientific' conclusion are facts until proven otherwise. Especially when the evidence is consistent with alternative hypothesis not allocating equal resources and consideration.

The multitude and composition of transitional fossils predicted and required by NDT were not found. Advances in science point to those that were found are more consistent with various ID and YeC hypotheses.

'Evolution Revolution' Dr. Spetner + 'Darwin's Doubt' by Dr. Meyer's show why the science precludes NDT. If NDT was possible, being a much greater claim, it would far greater burden of proof than the much more reasonable RCCF-YeC.

If anyone wants to remain willfully ignorant of this, that it their prerogative. Just realize that it is a faith based dogmatic belief in NDT, not sound science. If anyone holds by the big bang, start by a tiny area, but not the cosmic expansion / inflation components of SCM and SPIRAL, and holds nothing travels faster than standard light- speed, and holds visible starlight from 13B LY distance, is to claim the universe is at least 26B years old. As it would take 13B years, even at full light speed, for the matter to get there, and emit the light for the 13B LY and year light trip back. Plus any time from when the initial singularity first came to be, until it started to expand and for stellar formation.

A 26B+ year old universe is a vastly greater claim, thus having a vaster burden of proof than the 13.8B under SCM. Each instant over 6k rounded to date of a continuous viable universe adds to the claim.

To be a viable scientific model it would also need to not contradict empirical observations and sound logic like Olbers' Paradox. A dark night sky with distinct light trails from many stellar objects visible. So how can billions of stellar objects emitting light for billions of years align with that? Does Cosmic Expansion resolve for SCM?

SPIRAL's hyper-dense proto galactic formation, prior to cosmic inflation expansion, and a uniform departure point of visible light from stellar objects at, or beyond, the LY distance and years ago = to thousands, not billions, does.

Olbers' Paradox – Tackled

(+ see 'P-VoLT' on page 219)

Update: New Horizons 'the universe 'even darker than thought' best accounted for by far fewer galaxies, and/or the universe isn't nearly as old as current consensus thinks.[96]

All else being =, the older the universe is the less dark the night sky would be. So either the universe is not that old, and or there has been rapid expansion.

Ongoing Cosmic Expansion (OCE) would delay, and keep, a lot of starlight from being visible to us, depending on the distance and rate of OCE.

So if it is a static universe, it can't be nearly as old then if the universe is expanding. Even if OCE, is the night sky consistent with 13B years of light from stars formed within a 4B LY max?

Within YeC is the strongest resolution. If SPIRAL we predict and illustrate why aside from the light trails from CI, light has not reached beyond 5,777 LY to date (in 2016) if RCCF from any star/quasar.

SPIRAL provides the best explanation of higher ionization, brighter, galactic light, than expected if SCM.

Note how the objects are brighter, and more ionization than scientists expected. With no ready explanation under SCM-LCDM. Yet if SPIRAL, stellar objects started hyper-dense prior to cosmic inflation and did not reach mature density till the universe reached it's mature density, the more distant the object is now, the denser it was when it was 6k rounded LY distant.

So certainly crisper/brighter than if the light we see now from a galaxy half-way to the edge of the visible universe departed 6k years ago, from a distance of 6k LY, than if SCM 13B years ago starting from a distance of xB LY (still traveling 13B LY with the assumed OCE).[97]

---

[96]www.nasa.gov/feature/new-horizons-spacecraft-answers-question-how-dark-is-space
[97]https://phys.org/news/2019-05-clues-ancient-galaxies-lit-universe.html

Time lapse at a steady luminosity output, before considering Cosmic Expansion (CE) and a start to the universe, one would predict a bright as day night sky.[98]

SPIRAL provides by far the strongest explanation. SCM 13B Years / SPIRAL's 5778 = 2.25 M times more luminosity if SCM than if SPIRAL.

Now factor in CE that reduces the visible luminosity by elongating the frequency, The greater the CE velocity, the longer it takes for the fixed amount of output per fixed amount of time, to reach our field of vision, and the less scattered/shortened the frequency is.

The greater the stellar object recession rate, the less time it shed light at any point along its light trail.

Thus, with the change in density if SPIRAL it is easier to understand why a distant galaxy appears as a crisp single stellar object under SPIRAL than if SCM. Up to 730B c vs 3.27 c = 223B times more scattering if SCM than SPIRAL.

The light trails are also tighter when starting hyper-dense. At 2B LY radius divided by 5778 LY departure point = currently up to a factor of 346,140.53

If deep-time, much predicted light, is missing.

Dark energy, predicted and required by SCM ongoing cosmic expansion, is missing.

Just like the observations that cause SCM to predict and require dark matter may better be explained by the SPIRAL cosmological model, where one would not predict and require the missing dark matter so one would assume non-existent.   So too, the missing light that concerned Olber, and the missing dark energy required if SCM, align with, and help attest to, SPIRAL.

---

[98] http://math.ucr.edu/home/baez/physics/Relativity/GR/olbers.html

Another great advantage over SCM. If SPI-RALL one would predict the prevalent cosmological redshift of distant starlight and it's increase with distance due to cause and effect.

Rather than react to the observations of prevalent cosmological redshift of distant starlight and it's increase with distance, as does SCM..

'Brighter than scientists anticipated' compounds the dark night sky vs. opaque paradox for deep-time dependent cosmology models:

'NASA's Spitzer Space Telescope has revealed that some of the universe's earliest galaxies were brighter than expected. The excess light is a byproduct of the galaxies releasing incredibly high amounts of ionizing radiation. The finding offers clues to the cause of the Epoch of Reionization, a major cosmic event that transformed the universe from being mostly opaque to the brilliant starscape seen today.'.. 'The data show that in a few specific wavelengths of infrared light, the galaxies are considerably brighter than scientists anticipated. The study is the first to confirm this phenomenon for a large sampling of galaxies from this period, showing that these were not special cases of excessive brightness, but that even average galaxies present at that time were much brighter in these wavelengths than galaxies we see today.'[99]

Yet all galactic formation was about the same relative time. The solution may be SPIRAL where all galaxies beyond the inner universe we see are from a uniform years elapsed subsequent to cosmic inflation expansion that is = to the LY departure point distance, About 6k LY and years ago after all factors considered.

---

[99] New clues about how ancient galaxies lit up the universe
    https://phys.org/news/2019-05-clues-ancient-galaxies-lit-universe.html

Capped at 5M LY as by definition CR light was exposed as we know CR starts within that. Once one considers all the applicable science 6k rounded LY is in the sweet spot.

Overall the more distant a stellar object ended up at the end of cosmic inflation expansion day four, the denser they were as their visible light here and now departed. Thus, higher ionization and brighter. With no deep time, that would result in an even brighter, opaque, sky.

(Update - 'P-VoLT' Olbers' evaluator at Keating checklist point 3) So Olbers' Paradox aligns best with what SPIRAL predicts.

- Age: less time for light to scatter in all directions.
- Cosmic Expansion (CE): the same item that SCM relies on. The starlight is not visible here for as many years from each star as the current light year distance of each star, be it up to 46.5B if SCM, 4B if SPIRAL.
- 'Blue-shift Offset' (CB) the dense proto-stars/galaxies far sides CR was enhanced during CI. The greater the redshift the longer the wavelength the less scattering, and at some point beyond the visible spectrum, thus leaving our very crisp view.
- With SPIRAL all light beyond the IU/OU border are from lagging light 'LL' trails. Any rotation during the CI was in a tight trajectory. For the collimated light. Not orbital peculiarities. So like using high speed film vs a camera or object moving out of sync.

Both SCM and SPIRAL hold the natural observations indicate stars formed relatively early in history. Both hold all the matter the stars are composed of distributed rapidly, during CI epoch's 'same day shipping' :)

Per SPIRAL from the central production facility. In consolidated packaging. To provide us light. Signs for hours, days, nights, months, years and the age of the universe..[100]

Cal-Tech findings of light emission earlier than predicted under SCM, are more consistent with SPIRAL where it is possible to see light from the earliest and most distant stars. Showing science has a lot to learn.[101]

SCM also holds a star may not be where we now see its light emanating from. So SPIRAL's claim we see CR from where the star was, not is, is widely accepted. No law of science precludes stellar formation was in close proximity.

A problem if one holds SCM but not cosmic inflation expansion (CI) is light from stars over 12B LY distance in opposite directions if the universe is 13.8B years old with a light speed limit of c?

If one claims no CI and holds by the other tenants of the SCM, one is asserting the universe has to be over 13.8B years old. The older the universe the greater the claim, thus requiring greater evidence, to earn equal credence.

Based on scientific law regarding the conservation of matter there is a fixed amount of physical matter. So under SCM the universe is less than 0.01% as concentrated as it was by the end of the inflation epoch.

As the area of a sphere with a radius of 46.5B LY is about 1000 times that of a sphere with a radius of approximately 4.65B LY. The volume inside a sphere = 4/3 x 3.14 x the radius cubed.

[100] Gen. I:14 and Rashi on Gen. 24:55 'Yamim' equated to Year
[101] www.caltech.edu/news/farthest-galaxy-detected-47761

Alternatively, the fraction of a sphere to a cube it fits snug in: 3.14 divided by six = .5233. An easier way is to consider: a cube is 1/1000 the area of a cube with 10 x the diameter. Now reduce the cubes to spheres.

Using the same average density SCM requires far more matter in the universe than if SPIRAL. If no OCE the radius of the universe might be a lot smaller at 13.5B LY. Perhaps vastly smaller, 4B LY if the most distant visible galaxies under SCM were never subjected to CE post CI.

SPIRAL provides the mechanism, a day four cosmic expansion event that in theory could leave the most distant stars at well over 13B LY distance rounded. However, that distance for the most distant visible stars is based on current popular deep-time, ongoing CE, and parallax assumptions.

Using SPIRAL assumptions, the most distant visible starlight might turn out to be well under 4B LY distance. Just parallax alone might turn out to be inflated by 30,000/5,777 = 519% or more.

Our calculations assume the universe as a globe sphere, as does SCM for the observable universe. The spherical observable universe with that radius would require 1k times less and 1M times less matter respectively than one with a LY radius of 46.5B. For now suppose a 2B LY +/- radius.

We hope to be able to measure and agree on the current actual average density of matter within our universe, then apply to either Model using it's own assumptions.

Later we will see the implication of this. While SPIRAL is in line with the amount of actual observable matter/energy, SCM is dependent on the elusive dark matter.

After much effort, it may not just be missing but may be mostly non-existent. If so SCM is unworkable, and The SPIRAL would be the only choice between the two.

SPI-RALL estimate is the RALL caused by cosmic inflation started an estimated one +/- LY from us. Then receding one LY per annum, so it was a material rate of change to start.

Fifty percent year one AM, 33% year two, 25% in year three.., that the first person Adam would have noticed. As we know Adam was not just the first prophet of Hashem, but also the first true scientist, observing the cosmological phenomena, even making predictions based thereon.[102]

'..The Inner Heavens and the Outer Heavens..' [103] While Adam may have noticed a receding outer universe / growing inner universe. By revelation at Sinai the inner universe would extend 2448 LY, expanding just 1/2448 that year. It is safe to say an observer then would not notice that small fractional rate of change. Unless they knew to look for it, if they had a tradition going back to Adam, or via divine revelation.

Moses and later King Solomon, appear to be aware of the outer and inner universe. Did they, or do we yet, have the tools to use it for an independent precise year count? Either way, Solomon did have a precise 480 year count from the (2448 AM) Exodus till Holy Temple I.[104]

---

[102] Talmud Avodah Zarah 8a

[103] Deut. 10:14, I Kings 8:27, II Chronicles 6:18

[104] I Kings 6:1 See 'Legacy of Sinai' R' Z. Fendel + The TD Chronology

Life experience gives context. Context increases understanding if in conjunction with adequate consideration, sound premise and logic. So please weigh in how your perspective sheds light how/why SPIRAL is reasonable.

- Hyper-dense Proto-stars/galaxies preceded CI/CE.
- Denser so more mass per set area for stronger gravity.
- Combo of that gravity & CI formed disk.. galaxies..
- So may appear as if Black Holes at galactic centers.
- It is an Earth/Sun Ecliptic centric spherical universe.
- A relatively stable universe. As CR not due to OCE..
- Cosmological observations align with YeC.
- Cosmological observations align with ID.
- CR attests to SPIRAL's L (Lagging Light from CI)
- CB attests to very dense stars preceding CI 'SPI'
- Cosmic blue-shifted objects beyond the 'IU' are due to past, not ongoing, movement in our direction.
- The thousands age/distance traveled by visible distant starlight trails if SPIRAL explains our crisp view.

While peculiar velocities helped locate one repeller 'In the linear regime of gravitational instability, repellers are as abundant and dominant as attractors.'[105]

SPIRAL's CB may help explain these complex flows.

Denser proto-stellar formation expansion during CI may have offset the one LY per year increase in distance from distant starlight's departure point. As the further the object the smaller it appears, all else being equal.

---

[105] nature.com/articles/s41550-016-0036 Yehuda Hoffman PhD

'Electro-Magnetic Repulsion' on cosmic expansion (CE):

If SPIRAL, hyper-dense proto-galactic formation started as early as day one. Light couldn't yet escape. On a macro level think of each galaxy, a popcorn kernel. Micro-wave to expand-inflate them out rapidly. So, aka **'Jiffy-Pop'** hypothesis :).

As pointed out in HTP: Each hyper-dense proto galaxy relates to the formation of it's electromagnetic field. That relates to the CI expansion epoch. So, said fields approximate the number of galaxies. The CMB may be the residue evidence thereof. Smaller and more numerous fields are a better design.

After our hypothesis we came across some solid insight that helps corroborate SPIRAL's 'Magnetic repulsion'. Was the energy/fuel for all cosmic expansion mechanism 'magnetic repulsion'? We alleviate Nobel Laureate Adam Reiss concern if one has to invoke three mechanisms for 3 expansion epochs: 'Inflation, acceleration, anomalous'. If SPIRAL we need but one.

In the same interview Brian Keating PhD alerts us to 'Magnetic Fields exist in all structures that are 'gravitational bound". If SPIRAL all Galaxies, and the universe as a whole, are what we can define as so bound. As gravity and magnetic repulsion have been in equilibrium from the end of day 4.

So no ongoing cosmic expansion. So another SPIRAL benefit is there's no need to predict & require dark energy.[106]

'Primordial Magnetic fields', that SPIRAL's 'Electric-Magnetic Repulsion invokes as the mechanism used for CE, are Dr. Keating's favored explanation for 'Hubble Tension' (which may be resolved with SPIRAL's 'cosmic blue-shift offset').

The result: 5781 years ago, by the end of 4/365(5781) of the history of the universe, our stable, on the first crest steady-state oscillation (SSO), 'big-bang' into 'static' universe.[107]

The proto hyper-dense electro-magnetic field of each galaxy had the force for 'cosmic expansion' of itself and to repulse the galaxies apart. After expansion what is left now at each galactic center is past, not ongoing, hyper-density the illusion (look-back) of a once hyper-dense 'Black-hole'.

---

[106] Brian Keatin + Adam Riess PhDs 'A Nobel Mind on a Cosmic Quest' June 29 '20
[107] David Partovi PhD - stable on the first crest, that long, works with SSO.

Those magnetic fields should replace 'Dark Matter' as the agent explaining galactic rotations, as we see the Galaxies beyond 5781 LY when at 5781 LY, to date, from mid-hyper CE day 4, when denser, with density correlated to 'resting' distance.

'Magnetic Repulsion' hypothesis provides the lattice that may best account for CMB temperature distribution.[108]

Stars 'evolve/d' rapidly due to high mass is consensus. Per SPIRAL & SCM the entire universe, with all matter & energy of all galaxies, started hyper-dense. How much more so could mere individual proto-galaxies start hyper-dense?!

Think of our Heat, Time & Pressure variables. Vibrating strands of energy applied to the hyper-density. At frequencies to form & incubate proto-stars & galaxies w/ vastly greater pressure.

Is Magnetic Repulsion the cause & effect mechanism of cosmic expansion? While 2 + magnets now repulse each other into existing space, so, they move apart, and we can measure their speed. Prior to space, would not space result between them, so they stay put, while the metric of space expands/forms between them. See 'SPIRAL vs Hubble exhibit CE'.

The electromagnetic build and resulting repulsion force between the proto-galaxies and stars therein, could be the cause/process for/of the CI event. See how far smaller magnets can repulse each other. ' Our magnetic center' what one might expect in galaxies if SPIRAL 'Magnetic Repulsion.[109]

If SPIRAL we see all distant stellar objects (whose light reaching us here and now was ever subjected to cosmic expansion, from when the universe was smaller/denser), from the approximate distance we see CMB from, which is one LY distance per year elapsed, subsequent to the universe attaining mature size and density, by the end of cosmic expansion.

For example if that was at the end of literal day four, 5,781 years ago to date, we see all the distant galaxies, including CMB & 'black holes', from when at that LY distance, when denser.

---

[108] Cosmic Inflation Theory Faces Challenges, Anna Ijjas, Paul J. Steinhardt, Abraham Loeb, Scientific American, 5/5/2017

[109] https://apod.nasa.gov/apod/ap190619.html our mag. ctr. pic a day 6/19/2019

Overall, the more distant the resting distance, the denser, the earlier during CI day 4 they passed radius 'I'. Objects w/in radius 'I' 5781LY to date, we see at mature size & density, & their regular light that was never subjected to cosmic expansion.

Now relative size appears constant if density & distance proportional. For an example of just size and distance, think Solar eclipse. The moon being 400x smaller and closer, yet to an observer on Earth, the moon looks about the same size as the sun.

So an object 6M LY would appear to us the same size as the same object when it was 1k times denser at 6k LY distance. Also see our observations on Parallax.

Now we predict cosmic expansion was a type of cosmic inflation at it had to be fast enough to account for the near black-body CMB, and took us from not a small hyper-dense area all the way to attractive and repulsive force equilibrium when the universe reached mature size and density.

So 24 hours would suffice, perhaps half of that. We predict 4 days max. till the universe attains mature size and density.

Thus when we see distant stellar objects, from over the LY distance = to the number of years elapsed subsequent to the cosmic inflation/ cosmic expansion epoch, we predict 5781 to date (SPIRAL caps that at 5M years max., years = to the nearest stellar object we see here and now whose light has ever been subjected to cosmic expansion). It was at that distance but a fleeting instant, as it left it's light trail, during cosmic expansion. Overall the more distant an object, the faster it reseeded.

Now if we're seeing gravitational bound massive galaxy clusters from where they are now, if billions of years (keep in mind under cosmic expansion the galaxy does not move, adjust for orbital peculiarities, just the metric expansion of space between us), they'd appear all cloudy, or lit up. It would be like our trying to see individual stars during the daytime. Thus SPIRAL resolution for Olbers' paradox is not just that the stellar objects are thousands, not billions of years old, but that we see their light trails from a distance that they passed within seconds.

Imagine hearing a distinct sonic boom of a jet fighter, not the whole sonic boom 'trail' of many miles. The departure point proximity, recession velocity, and density factor, of distant stellar objects, enhance our viewing experience of their visible light trails, that formed during the cosmic inflation expansion epoch.

Magnetic Repulsion can account for the CMB Temperature. Based on the formula for the volume of the area of a sphere = 4/3 x pi x radius to the 3rd power.

Givens:
- CMB temperature is near black-body over the entire universe.
- The current CMB temp. across the entire universe is what we measure here now.
- The temp is inverse proportional to volume. If the volume increases the CMB temp. drops & visa-verse.
- Using the current CMB temp we know the increase all the way back to it's hyper-dense start. ie a 10-fold increase in radius results in a 1000-fold increase in area,1/10th a radius was1k times the temp..

If SCM-LCDM the universe is 'Flat on the surface' the visible universe is a sphere with about 46.5B LY radius an entire universe of a volume = to a sphere with 10x the radius of the visible universe is a conservative estimate, with 100x reasonable.

A minimum volume of 250x that of the observable universe. A maximum volume well over 500x that.

If SPIRAL the entire universe approximates the sphere that is visible universe.

Use an estimated radius 1B LY (mid-way between our minimum and maximum size estimates).

Our estimated max radius 4B LY is based on the current consensus maximum visible light departure distance. It's estimated min. radius 40M LY (based on approximate consensus size of the visible universe at 'recombination').

If my math is close: SPIRAL vs SCM entire universe volume, LY radius of sphere equivalent.

Minimum radius values: 40M vs area 10 times the volume of a sphere with a 46.5B LY radius = 1M (1.57) (10) = 15.7M times the volume if SCM vs SPIRAL

Upper volume values: 4B vs 46.5B (100).1k (1.57) (1M) = 1.57 Billion times the volume if SCM vs SPIRAL.

Centrist estimates radius values: 2B vs 46.5B (10) . 12.57(1000) (1M) = 12.57B times the volume if SCM vs SPIRAL.

4B vs 10x volume of sphere w/ radius 46.5 max SPIRAL vs minimum SCM 1k (1.57) (10) = 15,700 times the volume if SCM vs SPIRAL.

40M vs 46.5B (100) min. SPIRAL vs max SCM 1B(1.57)(1M) = 1.57 quadrillion times the volume if SCM vs SPIRAL.

As the current temp is a given, the CMB temp from it's hyper-dense start, until the radius size of the universe used for SPIRAL (where the entire universe approximates the visible universe), were these amounts times cooler if SPIRAL compared to SCM, at each value, along the way, as the universe expanded. With CMB being: x times cooler to start if SPIRAL compared to SCM.

A minimum 15,700 //.

A maximum 1.57 quadrillion //.

If centrist radius values used 12.57B //.

If minimum radius values used 15.7M //.

If upper radius values used 1.57B //.

Thus the relatively cool process of magnetic repulsion is a vastly more parsimonious explanation for cosmic expansion, and one would have a heavy burden of proof to claim SCM instead of SPIRAL.

Keep in mind an object appears the same size if it is 1M times the size and distance as a closer smaller one.. For example, the sun is 400 times the size and distance as the moon.

So yes, Magnetic Repulsion may account for all the expansion force/energy of cosmic inflation cosmic expansion, with all the empirical observations as is, if SPIRAL, where cosmic expansion is predicted to have ended within 4 days, within 4/365(5780) of the history of the universe, and the entire universe being a tiny fraction of the volume/size than it would have to be, to account for the same observations, if SCM.

As Dr. Keating suggested, Electromagnetic radiation may have been one of three forces for cosmic expansion. So they average 1/3 of the whole. If SCM 1/3rd of the whole is far more than 100% of SPIRAL! So Magnetic Repulsion may certainly account for all CE if SPIRAL. Regardless of which force/s caused the universe to expand, SPIRAL is a valid alternate model to SCM-LCDM.

The greater the heat, energy, mass, matter or.. all else being = the greater the claim. In science the greater the claim the greater the burden of proof. So the most parsimonious explanation of empirical observations should be the default. So SPIRAL should be the default.

Both SPIRAL and SCM hold: A light speed limit of standard light speed 'c' = 1LY distance per year.. CMB detectable here and now departure point from all directions = LY distant per year elapsed from early in history of the universe..

Neutron stars may have never been consensus less, and may have been even more, dense. See our 'Black-hole illusion resolution' ie no ongoing density by galactic centers. See neutron degeneracy pressure but don't assume a higher than neutron star density proto-star was/is doomed to collapse.

So if during the cosmic inflation expansion the average stellar object had the same density as a Neutron star, up to a quadrillion or so times the magnetic force of our magnetic field. With far greater potential If/when denser.

'Another type of neutron star is called a magnetar. In a typical neutron star, the magnetic field is trillions of times that of the Earth's magnetic field; however, in a magnetar, the magnetic field is another 1000 times stronger.' .. 'A magnetar called SGR 1806-20 had a burst where in one-tenth of a second it released more energy than the sun has emitted in the last 100,000 years! '[110]

We observe certain insects, fish, animal & plant, species, with many tiny eggs, live offspring or seeds, that disperse and grow rapidly. Just like mankind was designed for great diversity, so too animals, birds, fish, plants, minerals and stellar objects. As the galaxies and stars are not uniform, the amount of repulsion and attraction varied.

An analogy to hyper-rapid, hyper-dense proto-galactic dispersion is by the moment of human conception. When billions of zinc atoms disperse in a flash.

The more distant an object the smaller it appears all else being equal. The most distant stellar light departure points annual recession is 3LY If SCM vs 0 if SPIRAL.

---

[110] A thousand x trillions is quadrillion/s the magnetic force..
    https://imagine.gsfc.nasa.gov/science/objects/neutron_stars1.html

If OCE wouldn't the appearance of CR'ed objects diminish & change relative position in relation to those not subject to OCE?

That the constellations look as they did 3k years ago indicates no ongoing cosmic expansion (OCE)!

Variables: Velocity, Time, Cumulative CE distance. Change in Density,/Brightness.,Trajectory from observer..

So SPIRAL's 'Magnetic Repulsion' hypothesis may best reconcile the factual natural observations with a hyper 'cosmic inflation' expansion, relatively early in history.

If Magnetic Repulsion was the cause of CE it means the alleged Dark Energy is an illusion, based on the faulty assumption of OCE.

'Cosmological Blue-Shift Offset'

If SPIRAL the proto-galaxies and their stars were hyper-dense. OK if no light escaped, pre-Cosmic Inflation (CI) expansion epoch. During CI they expand to mature density. So cosmological Blue-Shift offset (CB) on the near side, and far side 'CR enhancement'. With galactic collision and x-ray binary boosters? All else being equal:

As Galaxies expanded, near side of each expanded back toward us. So the larger the galaxy the greater the CB. Overall the greater the distance a distant galaxy is from us, the greater it's overall cosmological redshift (CR).

CB offset potential overall diminishes with distance. Predict no CB within 5781LY to date. Any blue or red-shifted stellar objects within that 'inner universe' (IU) aren't cosmic expansion (CE) related. Past that radius 'I' the closer a galaxy, the more CB can outweigh CR.

As explained in the 'leavening raisin dough' analogy under SPIRAL the closer the galaxy the slower the rate of distancing caused by linear CE. In other words, the less cumulative cosmic expansion it was subjected to.

This aligns well with the 100 +/- known blue-shifted galaxies being in, or near, our local group. As CB can more than 'offset' but w/ distance, is eventually outweighed by CR.

On the macro universal level, the greater the distance from us the greater speed with linear expansion. Likewise, within each Galaxy the more distant from the center, the greater the expansion speed during CI. The individual stars expanded, causing galactic expansion, causing the universe to expand. Did expansion to mature density take the same for an individual star, galaxy and the universe as we assume?

'..Blue-shift Offset' may explain the 'Hubble Tension' of the 'sound horizon' / Hubble flow speed, not being a uniform linear increase w/ distance of distant stellar objects.

So simultaneous at 1:1. Perhaps over 12 hours. If not try values up to 24:1 with individual stellar density reached w/in one hour minimum and the universe at mature density w/in 24 hours. Consider values for a universe, if as low as a 50M? LY radius.

If SPIRAL one might predict the further from it's galactic center a stellar object on the far side of a galaxy (at least 5780 LY distant) the higher CR on average than it's closer companions. (just as more CB offset on near sides..). In some cases beyond the visible spectrum depending on distance from it's center, and to us.

The closer the galaxy whose visible light departed when still subjected to CI (LY distance greater than years elapsed subsequent to CI) will on average exhibit more CB (cosmic blue-shift offset) the further out back toward us on near side arms, and greater CR the further out away from us on far side arms.

If LCDM-SCM why are visible blue stars on near side arms of spiral Galaxies, rotating sideways or away from us, more blue than far side rotating toward us? SPIRAL's CB!

Blue-shift & redshift is also caused by regular orbital velocities coming & goings. All else being = the closer the galaxy the more noticeable. So negligible and overwhelmed by cosmic expansions residual effect in distant galaxies.

If SPIRAL, unlike LCDM-SCM, predict far side arms measure higher CR than a bit more distant galaxy near side arms. As near side arms were subject to CB offset..

The appearance of 'Black holes' at Galactic centers may attest to SPIRAL hyper-dense galactic formation prior to CI. If SPIRAL we predict reduced mass/density, by each galactic center, from whence came the matter, that are the stars, of each galaxy.

As we postulated above the closer to the center of each galaxy the less CB offset, so near side and nearer the center of each galaxy whose visible light is still subject to CI on departure, will measure higher CR so appear to those who think in consensus Black-hole context, give the illusion of being 'drawn' toward the black hole.

Whereas if SPIRAL, the reality is they aren't being drawn, just as there's no ongoing expansion causing CR/CB. Reference SPIRAL's 'Black-Hole Illusion Resolution'.

All else being equal the closer the Galaxy the more the blue-shift may have offset the CR. SPIRAL predicts we are not on course to collide with CB blue-shifted galaxies as CB is due to past, not ongoing, movement.

Olbers' Paradox is better solved by far side galactic CE-CR enhancement extending beyond the visible spectrum.

Keep in mind if SPIRAL there's no local CR and CB, inside a LY radius = years elapsed since the end of CI-CE.

If SPIRAL no CE subsequent to CI-CE epoch. Was CI and the universe's spacing due to magnetic-repulsion? Reference SPIRAL's 'Magnetic Repulsion' hypothesis.

'Culmination of 26 years of ESO observations of the heart of the Milky Way 26 July 2018'

The 'First Successful Test of Einstein's General Relativity Near Super-massive Black Hole' and/or confirmation of SPIRAL?

If SPIRAL the entire physical universe approximates the sphere that is the visible universe with a radius of up to about 2B LY radius. So in theory, if SPIRAL, there will be no more CR light in the entire universe w/in 2B years.

Yet long before that we will see the effects as regular light (never subjected to CE) from any object w/in x LY from any observation point, with x = to 5,780 to date, & increasing 1 per yr.

This might be confirming SPIRAL's 'Cosmological blue-shift offset' and 'Black-hole illusion resolution' and not consensus 'black-hole ongoing hyper density' which we argue is an illusion!

Note how the galactic center Black-hole, that is by far the nearest, ours, that we can examine best, is 'not very active' and that around it 'in a holding pattern'!

'Not very active' as perhaps not hyper-dense, 'in a holding pattern', perhaps since the it lost it's hyper density, which if SPIRAL was by the end of day four, after the CI epoch. When the modern density, size and orbital particularities were set.[111]

If you can see what's inside a black-hole it isn't so dense as to prevent escaping light. Just as the nearest galactic center black-hole may no longer be hyper-dense, and what is about it may have started inside it, so too with all the rest of them.

'Redshift Quantization' (Periodicity) may be evidence of, and reconciled by, SPIRAL's 'Cosmic Blue-shift offset'.[112]

With SPIRAL hyper-dense proto-galactic formation, if this stellar object was just getting to the point where it was coming toward (blue-shift offset) to the turning away (enhanced CR) it might appear like this by the time the light that departed about 5748 LY - 5778 LY reached us over the past 3 decades. (the object wasn't 20k-30k when the light departed but about 4 to 5 times more dense when closer). See SPIRAL vs Hubble.

So it doesn't confirm consensus black-hole theory, even if it is also consistent w/ that, as it is also consistent with SPIRAL.[113]

---

[111] https://apod.nasa.gov/apod/ap190619.html

[112] http://www.setterfield.org/Quant-Redshifts_and_ZPE.html

[113] www.eso.org/public/news/eso1825/

Just like cosmological redshift (CR) means lagging light, that takes longer to arrive than an equal time span of light emitted from a stellar object at a constant distance from us. Cosmic blue-shift means compressed light, that arrives over a shorter span, than had an equal time span of light emitted from the same object, had it remained at a stationary distance from us. Ie not been subject to cosmic expansion or net cosmic contraction.

'The stars present within a newborn cluster will typically run from the upper-left to the lower-right: the main sequence. As a cluster ages, the stars 'turn off' the main sequence, as the upper-left stars die first. Based on where this turn-off appears, we can date the age of the cluster. However, in all open and globular star clusters, a few blue stragglers, stars higher-up on the main sequence than normal, can be found.'[114]

Is it age & temperature or 'CB offset'! Also if SPIRAL we see from when 5781 LY and YA to date. The more distant the 'resting' orbital peculiarity the latter in the CI epoch we see it from the more dense, overall. 'Pearlman vs Hubble' 'exhibit CE' applies on the galactic level too. So further out on a galactic arm allowed more CE or CB velocity..

Chicken or Egg? What came first, 'Black-Holes' at Galactic centers or stellar formation? SPIRAL: hyper-dense proto-galactic formation, as density transforms into mature size and density, relatively early in history. Both proto-galaxies and base-Min proto-bird kinds, were created in full stature week one. Those birds had eggs. So bird then egg..[115]

[114] Christopher Tout, Nature 478, 331–332 (2011)
www.forbes.com/sites/startswithabang/2018/12/03/this-is-how-the-universe-makes-blue-stragglers-the-stars-that-shouldnt-exist/#7cdef70f514d
[115] Moshe Emes Recent Complex Creation Framework on 'Founder Effect'

SPIRAL's '**Black-hole Illusion Resolution**' Hypothesis.

If SPIRAL's hyper-dense proto-galactic formation, start. predict the appearance of 'Black-holes' at galactic centers So past, not ongoing, hyper-density. Just as SPIRAL explains Cosmological Redshift is evidence of past, not ongoing, cosmic expansion. The illusion: hyper-dense now.

The visible light we see now is from the number of years past, equal to the LY distance it traveled to get here. The galactic centers are where the matter, that now comprise those galaxies, came from. Assuming hyper-dense galactic formation preceded the galaxy reaching it's modern density.

If so the area claimed to be hyper-dense 'black holes' at galactic centers couldn't be as dense now, as each had been. As that area included any matter there now, plus most other matter in that entire galaxy. Adjust for any gains/loss to nearby galaxies, as we see them from when closer/denser.

So proto-stellar formation by galactic centers.[116]

Perhaps some proto-dense stars, or galaxies, only partly expanded, so are still hyper-dense 'black holes'?

'Wheel spoke' sync like rotation may be as we see galaxies from mid-hyper expansion day 4, think of a rotating leavening raisin dough ring danish. Here raisins are stars.

If stars move at like speeds in their respective galaxies. Hyper-dense proto galactic formation may account for this, rather than the consensus postulated Dark Matter (DM). If the missing DM is absent, it falsifies SCM-LCDM.

Not finding DM supports stellar relative velocity was set during a hyper-dense phase. So supports 'SPIRAL' & our 'Black-hole illusion resolution'.

While no longer hyper-dense, we see galaxies now from 5782LY to date mid-expansion. The greater the 'resting' distance, the denser they were, where we see them from.

---

[116] A nearby example? https://youtu.be/W-FIjFMIuzM Anton Petrov

Black holes aren't one-way consumers. 'a combination of gravitational and electromagnetic forces sprays most of the gas away from the black hole'. If they ever drew in vast amounts of what was around them, is speculation based on consensus. Keep in mind the empirical observations we do see is what came out, a relatively long time ago, that may have been there to begin with, so aligns with SPIRAL Solution: started hyper-dense.[117]

If SPIRAL we are seeing 'Black Holes' that are over the LY distance = to year lapse after the end of cosmic expansion that included cosmic inflation, when they were at that distance, on their way to their mature size and distance, by the end of cosmic expansion later that day. When the universe attained mature size and density and the stellar objects settled into their mature orbital peculiarities.

We predict that is 5782 years to date, so we see from when 5782 LY distant, when at a greater density proportional to the distance they attained, at the end of cosmic expansion.

Take SPIRAL's 'cosmic blue-shift offset' of near side galactic objects flip side, far-side SPIRAL 'cosmic redshift enhancement' that can extend beyond the visible light spectrum, when viewing distant stars and galaxies.

If SPIRAL's Black-hole illusion resolution obviates the current consensus view of 'singularities' by galactic centers where in theory the laws of physics break down. So no need for 'theory of everything' (ToE).[118]

Especially if even the initial 'singularity' was 'mustard-seed' sized, as described by RambaN.

New insight on Primordial Gas Cloud Turbulence, Element, BlackHole, Quasar and Galacy Formation.[119]

---

[117] Chandra X-Ray Observatory is in accord w/ SPIRAL : 'Cosmic Fountain Powered by Giant Black Hole' https://go.nasa.gov/2xhZu7r

[118] Brian Keating PhD, in at least one of his you-tube interviews in Sept. 2020.

[119] Latif,M.A.,et al.Turbulent cold flows gave birth to the first quasars. Nature 2022

Messier 87 (M87)

There Moshe Emes series for Torah and Science alignment are works in progress. So there are several exploratory sections in this book that I could emit or edit if time allowed. time volume II. I decided to comment on M87 after hearing Dr. Brian Keating interview Janna Levin.

Discovered in 1781 by Charles Messier, this galaxy is located 54 million light-years away from Earth in the constellation Virgo.

'Its most striking features are the blue jet near the center' – RMP: which may be evidence of SPIRAL's 'Cosmic Blue-Shit Offset' hypothesis as under 100M LY current consensus cosmic distance ladder is under the amount of cosmic expansion subjugation which increases w/ distance overall to overwhelm any blue-shift offset. Which is heaviest in near side of the larger stellar objects that, if SPIRAL, themselves expanded to mature size and density during cosmic inflation expansion.[120]

While the nearest 'black-hole' is consensus estimated at 26k LY at our Galactic center. Assume it has been 26k LY distant since the end of cosmic expansion that if SPIRAL ended 4/ 365(5780) into history, when the universe reached repulsion / attraction equilibrium, and mature size and density. Predict that was 5780 years ago.

So to date, we are seeing it from when it was 26k/5780 times denser & 5780 LY distant. When it was .2223 (17) = 3.78 sun widths wide, not 17. It is hard to view as a lot of static/dust, as a relatively high % is regular light, all from objects w/in 5780 LY, and only subjected to very low level of cosmic expansion, as 26kLY is relatively close. Orbital peculiarities motion factor into stellar visibility.

---

[120] www.nasa.gov/feature/goddard/2017/messier-87#.XxJRODJaaXZ.twitter

Now some of that 'static' may be residual from the many processes required not just for the Earth but all the elements required for our Galaxy.

Consensus has M87 relatively 55m/46.5B LY = 0.0012 the way to the end of the visible universe.

IF SPIRAL assume the universe is 2BLY, not 46.5B LY, radius. So M87 2.4 M LY distant. Yet we see from the light that departed it when 5782 LY distant to date, 5782 years ago, when proportionally smaller and denser.

Where: (Thank you to Neil Miller PhD for the edit)
dist.est = distance of M87 according to estimate = 55M LY
rad.est = radius of universe according to estimate = 46.5B LY
rad.SPIRAL = radius of universe according to SPIRAL = 2B LY
so dist.SPIRAL = dist.est * rad.SPIRAL / rad.est = 2.4M LY

That would mean it's about 23 times smaller. Now the mass could be the same, as the current consensus, as it was proportionally denser, 'Now the 'Black Hole' by M87 and that by our Galactic center appear about the same size from our view point'. If SPIRAL we see them from when they were the same distance from us. So if 26k LY holds up and 2.4M Then the mass of the M87 one is 92.3 time that of the one by our galactic center. Some things we learn from the amazing M87 are 'a Black-Hole can 'evolve' to a Pulsar'. A Pulsar is a Neutron Star with a strong magnetic field.[121]

Globular star clusters, dense bunches of hundreds of thousands of stars, have some of the oldest surviving stars in the universe. A new study of globular clusters outside our Milky Way Galaxy has found evidence that these hardy pioneers are more likely to form in dense areas, where star birth occurs at a rapid rate.

---

[121] Janna Levin PhD 'How the Universe got it's Spots' w/ Dr. Brian Keating 'into the impossible' series July 7, 2020

'M87, hosts a larger-than-predicted population of globular star clusters.'

'Evidence of M87's galactic cannibalism'
RMP - indicates closer together when 'stolen'. So helps corroborate SPIRAL.

"Studying globular star clusters is critical to understanding the early, intense star-forming episodes that mark galaxy formation. They are known to reside in all but the faintest of galaxies.' and "Star formation near the core of Virgo is very intense and occurs in a small volume over a short amount of time," [122]

RMP – helps corroborate SPIRAL 'HTP' hypothesis where 'Heat, Time and Pressure' are three variables in stellar formation, just like they can be/ are in Rock Formation. Thus increase the density /intensity, reduce the time, required for stellar formation.

In fact, we can't be sure stellar formation can even occur in accord with the competing consensus model, over deeper time, in phase 1-3, via accretion, (when the energy and matter of the universe was dissipated past the approx 40M LY radius assigned to 'Recombination'. Which if SPIRAL occurred at radius 40M/2.5M=16 LY. CMB temp. a given and area if SCM 2.5M times that of SPIRAL.

---

[122] The results appeared July 1, 2008 in The Astrophysical Journal.
Globular Clusters Tell Tale of Star Formation in Nearby Galaxy Metropolis
http://hubblesite.org/contents/news-releases/2008/news-2008-30

SPIRAL on Binary micro-quasar SS 433
and 'Manatee' nebula super-nova remnant shell W50[123]

High speed particle jets don't extend very far, and appear to have been formed rapidly. Within a few days. So that density & pressure factors are needed to make it add up.

Collimated light might best be explained by SPIRAL's hyper dense proto-galactic formation PRIOR to hyper cosmic expansion, as the light trails from distances over 'L' LY start to be subjected to cosmic expansion.

Estimated distance 15k LY.

Observations span 10 yrs., as of a few years ago.

Precession 162 days based on our vantage point.

That doesn't mean the 162 days (.44 of a year) between the actual 'heart-beat' emissions, if the portion of the light trail was emitted during hyper-expansion. If 24X60x60= 86,400 seconds of Hyper-Expansion / 15,000 = 5.76 seconds average per LY. So .44 of a LY formed in 5.76 seconds is every 2.5 seconds, not every 162 days. 162 days the time each 2.5 seconds worth of the light trail, during its formation, takes to reach here.

We are seeing the object now from when it was 5,781 LY distant. The number of years elapsed since that Hyper cosmic expansion epoch. 5781/15,000 = 38.54%

Gama Ray Heartbeat consensus spread over about 100 LY (3D) = a volume of 100x100x100= 1M square LY based on a light departure point of where it is now at 15,000 light years.

[123] https://apod.nasa.gov/apod/ap200831.html
SPIRAL's 'Jiffy pop' unlike larger twin SS 433 'kernel' never fully 'popped'.
www.aanda.org/articles/aa/full_html/2018/09/aa32488-17/aa32488-17.html
www.desy.de/news/news_search/index_eng.html?openDirectAnchor=1887&two_colu mns=1

If SPIRAL we're seeing it when it was at 38.54% of that distance and density. 100LY (.3854) =38.54 cubed =57,245 sq LY.

1,000,000/57,245 = .057 = heartbeat reaches 5.7% the area at 5,781 LY than it would have to at 15,000 LY. Such are being 94.3% = 17.43 times greater, than at 5781 LY.

Note how the particle jets appear to slow down with distance or at the least go from red-shifted to yellow. This may attest to how electro-magnetic radiation could have been the repulsive force for hyper cosmic expansion of the hyper-dense proto-galxies theorized in SPIRAL.

Either way, if the visible light of SS-433 and W50 had been subjected to that hyper cosmic expansion, as indicated by the distribution area of the heart-beat, that fits a smaller area better, it nears our prediction of 'I' as 5781 LY.[124]

Quasar JO313-1806, GN-Z11, PSO J0309+27,.. early complex stellar formation, toward the end of the visible universe, whose light departed when a lot closer.

So hard to argue lower entropy due to lack of complexity over a material portion of history. Other than cumulative entropy, why assume isotope decay rates... any different in space now than after 10% of history?

More complexity, earlier, than predicted/anticipated under SCM accretion and several generations, till certain stars,.., but no surprise under "RCCF' 'SPIRAL' and 'HTP' hypotheses.[125]

---

[124] 'I' is the radius of Inner Universe (IU), visible Light from stellar objects therein was never subjected to Cosmic Expansion (CE). So not CR'ed or CB'ed. LY radius = to yr. count post Cosmic Inflation (CI).

[125] GN-ZI1, PSO J0309+27 and J0313-1806:
www.nasa.gov/feature/goddard/2016/hubble-team-breaks-cosmic-distance-record
www.aanda.org/articles/aa/full_html/2020/11/aa39458-20/aa39458-20.html
https://public.nrao.edu/news/quasar-new-distance-record/

SPIRAL Disclosures:

- Variables in Stellar Formation: Heat, Time, Pressure..
- Proto-galactic density: SPIRAL-high, SCM-low.
- Stars (all?) formed in (clusters) as parts of Galaxies.
- Light speed is limited to the speed of light 'c'.
- Light is bent by gravity. Hyper-density may 'capture' light, as in black hole/singularity theory.[126]
- Light rebounds to 'c' if the impediment is removed.
- Light is not subject to inertia.
- Sound and other matter is subject to inertia.
- Doppler concept of redshifted light.[127]
- See 'Crest' & 'SPIRAL vs Hubble' for CR via CE.
- Red-shifted light's wavelength is longer, blue shorter.
- Blue higher oscillation frequency, Red-shifted lower.
- Higher frequency/shorter wavelength – light is more scattered
- Longer wavelength / shorter frequency for less light scattering.
- Resistance such as water can increase light frequency.
- CR is due to SPI so past CE, not ongoing CE (OCE).
- Blue-shifted galaxies due to CB, not ongoing motion?
- CB as dense proto-stars/galaxies pre Cosmic Inflation (CI)
- Peculiar Motions the closer the object the more effect.
- Universe 2BLY radius. Mature density & size by end of CI
- Entropy: time works against closed systems lasting.
- The physical universe is considered a closed system.
- If something can last a long time, it can a short time.
- Existed a short time, does not prove long lasting.
- Lasting a short time is more probable than long.
- CMB radiation indicates hyper-dense start, then CI.

---

[126] Jacob D. Bekenstein, Black Holes: Physics and Astrophysics - Stellar-mass, supermassive and primordial black holes. arXiv:astro-ph/0407560 (Submitted on 27 Jul 2004)

[127] Doppler Effect http://en.wikipedia.org/wiki/Doppler_effect 'For..light..only the relative difference in velocity between the observer & the source need to be considered.'

Additional Discussion:

Under SCM narrative, CE has been going on for up to 13.8 Billion (B) years averaging about three times the speed of light, since the end of the Inflation Epoch. That there is pervasive CR of distant starlight SCM assumes CE is due to ongoing cosmic expansion and the expansion has continued since those stars were formed. (SPIRAL holds no OCE.)

In theory visible light from stationary stars the number of years old equal to the number of LY away, should arrive as regular light with no CR and no blue-shift.

Blue-shifted galaxies may not be moving toward us, as what we see now is from in the past. The number of years equal to the LY distance the visible light reaching us now departed the object. As most stars formed early in history, that of the hundred of so within our local group have not collided and or passed us by now, is evidence it was due to past, not ongoing motion toward us. Just like SPIRAL teaches CR due to past but not ongoing cosmic expansion.

SCM like SPIRAL holds light speed is limited to C the speed of light. Therefore there can't be visible light emanating from beyond the LY distance equal to the age in years of the universe.

The 'visible universe' is the sphere we're by the center of that includes up to where the stars are now whose light we see now. Consider the Sun-light that departed just over 8 minutes ago. Were both moving. So we aren't seeing the Sun exactly where it is now. We are also no longer where we were when the sunlight we see now, departed the Sun. Now instead of an 8 minute, imagine 5777 year and 13.8B year delays. SCM holds a 46.5B and SPIRAL 2B+/- LY radius visible universe. So while we can't see beyond the LY distance = the age of the universe, we do see light from stars that are now well beyond that LY distance.

A static universe with no past or present CE claims more, but answers less, than even SCM. How did it get here? Entropy, CMB blackbody, CR, size/shape/density of the universe, dark night sky/Olbers' paradox... SCM and SPIRAL both hold the most distant visible stars formed within 4B LY.

Per SPIRALL stars beyond the LY distance = to years elapsed post CI we see from when they were at that distance. Per RCCF the end on the inflation day epoch was 5777 years ago to date. So we see stars/galaxies that are over 5,777 LY from when they were 5,777 LY distance while moving away from us. Don't forget to factor in CB offset.

So while in theory even with SPIRAL the CE-CI event could have been to 13B LY or even 46.5B LY, with no need for deep-time, 4B LY is the upper limit to radius of the visible universe's light departure point. At greater distances the cosmic distance ladder could be inflated well over two times. So 4B/2 = 2B LY is a reasonable radius of the visible universe based on the SPIRAL explanation of CR.

Under SCM CR is attributed to CE. The amount of CR has SCM estimate stars that started 4BLY distant were recently 46.5B LY distance from us and still receding.

Under SPIRALL after CI no subsequent material CE. So stars at the end of the inflation epoch should still be at approximately the same mean distance. Adjust for normal orbits and allow for a normal amount of aberrations, like out of orbit stars.

SCM like SPIRALL presume light from early, distant stars, traveled at light speed to reach us. Those with CR were moving away from us when that light was emitted. Those blue-shifted Galaxies in our local group had a motion toward us when that light was emitted. Disputed is if ongoing.

If it is true that the universe has been expanding over deep-time the ratio of stars whose orbits we can observe directly moving away from us should be greater than those moving toward us. If it is a fact it is about even, that should falsify that expansion is going on now.

If expansion has not been going on over deep-time, and or if the universe is relatively static, SPIRALL should be the most reasonable explanation of the CR observations, thus the best science.

Evidence of fast moving stars. If they are still in the very galaxy they were formed in, challenges deep-time and help corroborate RCCF-YeC and SPIRALL. As why have they not made it out by now if deep-time were true and most stars formed near the get go. See ICR.org and The RCCF for additional references.

Massive clusters in very deep space would not be predicted under deep time models, as how did all that energy/matter get out and coalesce out there?

Good with SPI of SPI-RAL as Galactic formation Preceded CI, where all energy and matter were in close proximity during the formative stage of those massive clusters. So they could have varied in size to begin with, or joined before CI-CE left them in deep space.

Under both SCM and Pearlman SPIRALL models If we can see light from stars now 13B LY away, they formed, and were already giving off light, relatively early in the formation of the universe.

Both models concur about relatively early star formation. SCM holds CR is due to an expanding universe, stars that were 4B LY away shortly after a cosmic inflationary epoch are now about 46.5B LY away. So SCM holds some stars formed up to 42.5B LY closer to Earth than the same stars are now.

SPIRAL would hold if any visible stars are 13B LY distant they formed near here and their light departed nearly 13B LY closer to us than they are now. The greater the claim, the greater the burden of proof. 2B LY is a lot smaller claim than 46.5-4=42.5B LY. SCM's 42.5B is 2,125% greater than SPIRAL's 2B LY estimate to the most distant visible stellar objects vs their lights departure point. Either way, they had to form early and much closer, SCM and SPIRAL agree.

The SCM holds the CR we are seeing now from the outermost visible stars traveled 13B LY in 13B years, before reaching us. A far greater claim requiring far greater evidence than SPIRAL's 5778 LY, 5778 years ago to date.

SPIRAL provides science a tremendous measurement tool, based entirely on cosmic observations. It provides how to precisely measure the age of the universe. The number of years elapsed from the start of the inflation epoch till the present, should be a direct one to one correlation with the LY number from Earth to the most distant stars that we get regular, not CR and Blue-Shift offset light from.

It should be a bit past the inner edge of the outer universe, where CR and Blue-shift offset begin. The more distant stars/Galaxies the more the CR outweighs the blue-shift offset. Each year on average only a tiny fraction: one divided by the radius of the visible universe in LY of stars regular light will begin to reach us.

So it may take many years of observations for enough stars to consider what is going on at the border as a 1 LY deep of stars lose their CR and CB as their normal light begins to reach us, so the radius of our 'inner' universe of no CR, and no CB, on departure, expands 1 LY per year.

So that border to the nearest CR and CB light departure point is an independent way to measure to 5778 LY to date and help calibrate the cosmic distance ladder.

That radius in LY from Earth to visible CR/CB light's departure point being equal to the number of years elapsed from the end of the inflation epoch. If SPIRAL a far greater proportion of CR/CB stars from 'O' reach us then regular light, from 'I' stars. So how do we find the tiny proportion of regular light from stars even 30% within the " I " boundary?

One variable to consider is distance, as the greater the distance the object the smaller it will appear all else being equal. It is thought we do not see most individual distant stars but the galaxies they are in. The number of individual stars we on Earth can distinguish by eye are in the thousands. Consider how light 'blends' so what looks to the naked eye as a star may be a galaxy..

Another variable to consider is the distribution of matter in the universe. Assume the Cosmological Principle's (CP) position of an even distribution of matter over large distances. Spiral would falsify (render invalid) the CP notion of a uniform view, see exhibits A-D.

With an even distribution the amount of matter in a distant 100 LY ring from can be 276 million times more than in the 100 LY ring closest to us.

As a reference point assume the big dipper is an average 86 LY from us. Add the seven stars LY estimates: 124,101,84,81,79,78 and 58 then divide by seven.

It is easier under SPIRALL to answer than SCM. Assuming a somewhat even distribution of matter. With that matter from our central distribution point during the CI day epoch, leaving the CR / CB light trails. Then after factoring for CB the further the light emitting object by the end of the CI, the shorter oscillation frequency so longer wavelength.

As the same approximate amount of light omitted from all large sectors that day, means that the more distant the sector, the more CR 'stretched' its light-trails wavelength.

It should be over 200 million times the matter is in the form of 200 million as many stars, the same amount of stars with 200 million times as much matter on average, or a composite of the two.

To check the math take the area of a sphere with a radius of 13B LY minus the area of a sphere with a radius 100 LY shorter. Then divide the result by the area of a sphere with a radius of just 100 LY.

As long as a consistent valid measurement system is used SPIRAL predicts the inner universe to expand an average of one LY per year.

To get the exact age in whole years of the universe determine the exact LY distance from the next batch of visible static stars, whose light is now regular, instead of red shifted, in the band one LY further from us each year, that per SPIRAL started receding from us at the end of the inflation epoch by one LY per year.

Then subtract the LY distance of the start point from which that light emitting object cosmic inflated away from us. SPIRAL estimates that distance at one LY.

SPIRAL makes a testable prediction of the LY distance to where pervasive CR/CB begins and that that distance increases one LY annually. The span of years has to be long enough to be noticed, which will depend on the method and technology.

58 years would add 1% if 5,777 to date. Falsifiable testable predictions is a good feature to achieve scientific theory status. As always use consistent definitions, be consistent in the use of measurement tools, and fully disclose: assumptions, methods and all the results.

SPIRAL can help affirm distant standard candles benchmarks. If RCCF the age of the universe is 5,777 to date, we know how distant a star is when it goes from CR or CB to regular light at the edge of the inner universe.

If SPIRAL we know the minimum distances for objects whose visible light is still CR / CB and maximum for those whose light is regular.

If SPIRAL there is no Hubble Volume/Sphere boundary where stars outside same will never be visible. The visible universe approximates the entire universe at a 2B+/- LY stable radius rather than 46B LY and growing if SCM.

If not 30k, but 5,777 to date, max LY distance departure point of visible starlight, this new standard candle benchmark by/for parallax triangulation will greatly help make a more precise cosmic distance ladder.

In The 'RCCF' we reference how radiometric dating, can have wildly divergent results, even when using the same rock of a certain age for multiple samples. Yet current popular science values as if a reliable deep-time factor.

SPIRALL should hold up a lot better under scientific rigor than radiometric rock dating would, if given the same consideration. The RCCF includes how to more accurately calibrate valid radiometric dating techniques to help maximize alignment of the results with the actuality.

Laws of nature are part of the design and creative process. Think of vibrating stands of energy, chemistry and physics, as the software and hardware.
See Moshe Emes vol. I & III for details & references.

- Year of Tohu – Hashem conceived creation's design.
- Day 1 - Hashem created all Physical Matter ex-nihlo.
- Days 1-6 - Hashem organized the physical matter.
- Day 2 – Hashem separated the matter above the sphere and the matter below/inside the sphere, which included or was Earth.
- The Earth was protected via electromagnetic radiation, in the 'eye of the hurricane' most likely a sphere or disc encircled the gap that circled the Earth.

It may help to think of the journey of a human egg, surviving and thriving, inside what would normally be a very hostile mass, vastly greater then its initial size, if not in just the right place, and if not for an amazing array of features in the host mother.

Think of the anthropic principle of how we are still in the only proven protected human life sustaining sweet spot of the entire universe.

Did the gap occur inside what already existed, or did the above sphere move out from inside the existing sphere? The initial distance between the earth and the outer sphere? The depth of the outer sphere? was it expanding and if so at what rate? to be determined.

Not every star is same. (likely no two stars are exactly alike). So the light they emit can vary. Too much elongation or shortening puts light outside the visible spectrum.

Redshift frequency shortening may result from:

- Redshift from a light emitting object moving away from us. With CE this resulted in CR.
- Redshift if a star is orbited away from us.
- Blue-shift if star moving toward us outdoes any CE.
- Resistance increases frequency shorten wavelength.
- Gravitational bounding could have offset some CI/CE
- Other Doppler type causes may be mistaken for CR

In these I suspect no real frequency shift:

- Brightened by aligned stars, dims if such an alignment ends. (Log in the fireplace analogy?)
- Dims if obstructions eclipse, brighten as any blockage clears.
- Dims if gravity slows, then brightens back if gravitation pull released. ..

The R in SPIRAL is for Cosmological Redshift (CR). SCM attributes CR to Ongoing Cosmic Expansion (OCE).

Per SPIRAL CR is due to past, not ongoing, CE. From when stellar objects moved away from us the observer. Either due to the objects relative movement away from us, or the cosmic expansion of the space in-between us.

At cumulative speeds up to greater than 'c' standard light speed. The same amount of light emitted per any set amount of time reaches us over a longer time span than if the distance between us did not increase during that departure span. The light thus 'lags' so gets redshifted in the direction the stellar object is moving away and/or receding from.

So to an unobstructed stationary observer, viewing the light trail from the point of departure of the receding light producing object, the trail is CR'ed. If the receding object stops emitting any light, the CR light will continue to appear to that observer for the number of years equal to the LY distance it was when it stopped giving any light.

If the object becomes static but continues to give light, after the same number of years as the LY distance from the stationary observer to where it became static, the light will change from CR light to regular (non-CR) light.

That happened on 'inflation day', day four, when the stars were formed in a much more concentrated universe. Then what happened to the upper layer from whose energy and matter the celestial bodies are comprised expanded like a microwaved popcorn, with cosmic inflation at speeds vastly greater than the speed of light, as touched on in 'inflation theory' expanding the universe to its approximate current static state within a day +/-.

I liked to call this The Inflation Day 'Jiffy Pop' microwave popcorn analogy expansion hypothesis. Now 'magnetic repulsion' hypothesis, to help us envision the near blackbody CMB background & widespread distribution of galaxies in the universe.

Now think of the universe at the end of the inflation epoch as a tent filled with galaxies, instead of a microwaved bag filled with popcorn. Each 'kernel' it's being it's own electro-magnetic field from early in CI should be much higher probability science.

That CI-CE post Hyper-dense Galactic formation may explain how and why the light trails from distant stellar objects were focused & subjected to CR (see Pearlman vs Hubble).

It also explains the Galactic rotation observations (see SPIRAL 'GriP') So unlike if SCM, the missing dark energy & dark matter are neither predicted or required.

Helping corroborate SPIRAL is the appearance of past hyper-density by each Galactic center. Misconstrued as black-holes we are seeing evidence of past, not ongoing, density that preceded the mature galaxies.

SCM has no answer if the alleged 'black holes' came first or not. Just like creation in full stature bird base kinds came first, then laid eggs, so too hyper-dense proto-Galactic formation came first before their mature stellar distributions.

Proto-stellar objects emitting light moved away from us at speeds where their trajectory should be straight. The light trail they omitted left a CR trail. Redshift is the elongated visible light frequency so scatters the least compared to blue that scatters the most.

So Olbers' paradox fits SPIRAL the best as much less time elapsed thus less background light pollution along with prevalent CR so less scattering than if normal light. The exhibits A-D help illustrate why ours is the optimal view, with no time for distant stars to fog up the night sky. The longer the light has been traveling the more dissipation.

The greater the mass of the light, the more the light will be bent by gravity. So light that started out normal, or on a star moving away at less than C, will be bent by gravity more than light produced by a star moving away faster than C, all else being equal.

So, he observable evidence indicates: the Earth was in the approximate center of the CI-CE, the formation of visible distant stars preceded that inflation, and the Earth is the approximate center of the universe.

If SPIRAL, after cosmic inflation ended, so did CE. CR, after calibration for blue-shift offset, started at about one LY away and has receded at one LY per year since.

So, a reasonable assumption is over long distances the number of stars moving toward any point should approximate those moving away if CR is due to SPIRAL so not evidence of ongoing CE.

SPIRAL predicts if we can go straight, starting from the Earth ecliptic, passing beyond the orbit of any star, by a LY distance one or more greater than the age of the universe, we would observe that that stars light has yet to reach that point. That far side of any star LY distance, that their light has not yet reached, as predicted by SPIRAL under RCCF would be near 5778 LY to date. This assumes we have a way to get an accurate exact measurement, if not go additional LY's beyond that point to account for the measurement techniques margin of error.

So the typical stars light reaches us because it originated close to here, 5,778 years ago to date, some are perhaps younger, none are older. So any light on the far side from us of any star should not be CR due to CE, but should appear as regular light to the far side observer., and their light would not have traveled more than 5778 LY to date further away from us than their most distant position.

Some think of the universe in layers, and use an onion as an analogy. I am not sure how uniform the layers are, but suspect each 'row' is made up of individual 'blocks'. Think of the patterns one can use when making brick patio, Even better a stone wall with stones of various shapes and sizes.

With the SPI of SPIRALL each galaxy started out very dense. Like a group of billions of frog eggs floating on the circumference of a globe, each egg would expand to average x light years in diameter and group in discs/galaxies.

If there are 1000 layers, and it was 13B LY to the outer edge of the universe after the inflation epoch, then each layer would average 13M LY thick if a somewhat static universe and 46M LY thick per SCM today after CE. As a reference the milky way is estimated 100k LY across.

The popcorn as the individual kernels expand in no particular order, to the size of an expanded kernel, but some kernels get moved many kernel lengths while others do not travel as far, yet regardless of speed and distance traveled they reach their approximate destination at approximately the same time, which is in proximity to the end of the microwave induced inflation. The accelerated speed at which the kernels pop when microwaved compared to conventional stove top heating is another lesson.

Another popcorn analogy, not all kernels get popped just like not all cosmic matter turned into stars. Finally microwave popcorn can remind us of the cosmic microwave background (CMB). CMB measurements are scientific evidence for inflation theory with CE at speeds vastly greater than Light speed c.[128]

[128] Alan H. Guth PhD. 'The Inflationary Universe: The Quest for a New Theory of Cosmic Origins.' 1997

Recall the 'I' in SPIRALL is for CI. Observations claim to confirm cosmic inflation (CI).[129]

Additional analogies to microwave popcorn, human eggs, frog eggs, tuna eggs, plants, bubbles in a bubble bath, cluster spiral pyrotechnics. Hatched sea turtles and baby Tuna grow large moving great distances via sea-currents...

Bubbles sometimes merge with another so in a globe comprised of bubbles, even if each bubble started out fairly uniform, within a short period we could have super bubbles.

Now if each bubble had enough for one galaxy to start, the mass in multiple bubbles merging when they were in closer proximity may explain how super galaxy clusters and empty spans came to be.

Leavening raisin dough and bubbles are more common conventional analogies to describe cosmic expansion and the structure of the universe.

See NASA WMAP[130] for current conventional understandings, many of which should be revised based on SPIRAL understanding of the observations.

The 'Horizon Problem' of shared characteristics between distant sectors should be resolved if all the Galaxies were in close proximity to their neighbors, and formed about the same time, as in SPIRAL

The elusive 'Dark Matter' issue: consistent rotation of spiral galaxies beyond a certain point is understandable, as is stellar velocity and rotation, w/in individual galaxies even absent missing dark matter, if they formed hyper-dense at about the same time, and we observe those beyond the 'inner universe' from the same light departure point 6k LY rounded.

---

[129] John Kovac, Jamie Bock, Chao-Lin Kuo, .. astrophysicist BICEP team
http://bicepkeck.org/b2_respap_arxiv_v1.pdf
[130] http://map.gsfc.nasa.gov/universe/WMAP_Universe.pdf

Where CR that had been attributed to CE under the SCM can now be understood in the context of SPIRALL, be they from galaxies now 6k to billions of LY distance. As this year we are seeing them from the light departure point on that section of their light trail left on cosmic inflation day four, 5779 years ago to date, which would now be 5779 LY from us.

If SPIRAL the missing matter/energy theorized under SCM may be a case of 'Emperor's new clothes' so non-existent. Recalculate the big bang for this may better account for galaxy clusters properties and their orbital velocities.

Complexity earlier then predicted under uniformitarian assumptions is the norm and points to ID and YeC. Just as most of the great pyramid building taking place over a relatively short period of time, about 200 years, coming to an abrupt stop by the Exodus, is a clue that helps confirm the Torah testimony, so too the evidence that most stars formed over a relatively short span of time early in history is a clue supporting Torah testimony regarding creation week.[131]

One who understood SCM's issues, may already see how/why SPIRAL adds up to solve the puzzle. Like Pharaoh when he saw the truth in Joseph's explanation.

Joseph, aka Imehotep designed the first great pyramid, by/as a central grain distribution facility, during the seven year famine. Joseph used Torah Revelation (The word of Hashem) to reveal what was hidden 'Tsapnat Penah'.[132]

What all considered correct in hindsight, that which the world class academics of Mizrayim could not, despite their superior status, and resources. Let's save our world, as Joseph did for Mizrayim by connecting to, and not obstructing the wisdom in Torah, the true word of Hashem

---

[131] See Moshe Emes RCCF framework + 'Torah Discovery Chronology'
[132] Rashi on Genesis 41:45 Tsaphon means Hidden as well as North.

Deep-time narratives need frequent adjustment as advances in science bring in new data. Torah testimony was right to begin with, so it never needed revision. It is a consistent narrative that aligns with all known facts. A practical guide to reduce waste and advance science.

The best science being the best explanation of observations. Incidental is how that explanation came about.

Observational analogies, key words, and topics to visit: Snow, fog, flashlight, Jet plane steam trail, Traffic flow.

Slinky, inch worm pattern, Quantum Mechanical inflation Pattern. Sound waves direction/focus. Sound unlike light, subject to inertia. So can move at or above speed of light w/ object traveling greater than the speed of light.

Sound with no atmosphere? Heat, time and pressure. Volcanic action rock? Or magma? heats up at high temp, turns to liquid and gas, moves quickly, can travel far, lava cools and hardens back to igneous rock. Extrapolation. Central assembly plant by the power grid. fireworks and cluster bombs. Human, eggs in female.

Salmon congregate to breed, the next generation return to the ocean. Seeing the mature Salmon at sea should not make us conclude they hatched at sea.

Snow flakes start in a cloud then spreads out between the cloud and the ground, if we could only see a snapshot of the snowflakes but not the cloud, would we erroneously conclude each drop formed in its approximate current position? The clouds themselves may have formed over bodies of water then migrated w/ the wind.

Tree and flower pollen formed on plants and trees then migrated with wind or pollinating agent..?

On a grape cluster tiny grapes start to appear and expand along with the stems. I assume the same for pomegranate and some other fruits and vegetables..

Monarch Butterflies, Albatrosses, Penguins, Sea Turtles, Some whales and sharks,.. use central breeding and or incubation locations, then migrate/disperse. So we see plenty of patterns in nature that are consistent with SPI star formation prior to the day four cosmic inflation (CI).

We assume CI started day four by a 1LY radius, but some CE likely started as early as day one.

Chestnut trees: the many buds>blossoms>fruit w/ spiked shell, packing and core growing from a tiny unripe seed to a strong egg yolk sized nut. We honed our passing skills using a football to collect Horse Chestnuts. They have a common designer/creator as stars. What is the ratio of doubles, triples and the rare quadruple, in each?

Smoking gun evidence that Cosmological Redshift Attests to Lagging Light (RALL) from when the fledgling galaxies were incubated close by and surrounding the Earth.

Close by may mean a light year +- and may account for the missing matter in the asteroid belts and OORT cloud. This may also help explain the solar and star dust results by Dr. J Kissel's team at NASA.

Two more issues with SCM and the visible universe: How did any CR light much over the current 1.69 z (CR redshift) unreachable limit arrived here to begin with?

Why would the future visibility limit of 19,000MPC (3.26LY per Parsec) be much higher early in history then now?[133]

Neither a problem under SPIRAL as stellar formation was in close proximity to us. Also with no OCE distant light heading our way gets here in the number of years equal to the LY distance to go.

'Blending' of light − It is obvious a distant galaxy that has x stars may appear to the observer as a single star. That galaxy is rotating and or has peculiar velocities were some of the light should be red-shifted and some blue-shifted.

---

[133] www.astro.princeton.edu/universe/ms.pdf universe map

Yet from our vantage point that may be indistinguishable. Even local rotating stars starlight may 'blend' the portion moving away from us with light from the portion moving toward us.

As stars/galaxies expanded during CI, local CB offset might overwhelm CR. On the far side CR was enhanced.

SPIRAL may help prove a stable steady state universe about 5780 years old and be more consistent with the:

- Galaxy and Stellar Formation with prior to CI energy.
- Cosmological Redshift. As one would predict if SPI.
- Absence of the missing Dark Matter as we would predict if SPI Galactic formation 'just' 6k vs 12B YA.
- Absence of the missing Dark Energy, if no OCE.
- Earth being the approximate center of the universe.
- The optimal viewpoint of the universe is from Earth.

Anomalies like Lithium measurement, challenge SCM more than SPIRAL, just like genetic information challenges NDT Darwinism more than it does Creation Science.[134] [135]

SPIRAL is a far lesser claim than Baryogenesis, Big Bang Nucleosynthesis, and an Era of Recombination in deep-space, as hypothesized under SCM.[136]

While SPIRAL is an acronym unrelated to the definition of the word spiral, we played football. Think of how a spiral pass can increase accuracy, distance, separation and speed. Like a riffled gun barrel. David slung a smooth stone. Like w/ rock skipping. So perhaps the how of galactic formation and placement by SPI-RALL had some spiral :)

The 5 W's: Who – Hashem   What – The Physical Universe. When – 5,780 years ago to date. Where – Starting from relatively close to here. Why – For us. (to exercise our free will to be aware and act accordingly in order to grow our connection to The One Heavenly Father and each other). How? Learn and consider Torah and Science with awe of Hashem.

---

[134] 'Big Bang Fizzles under Lithium Test' Brian Thomas, M.S. ICR 09/22/2014
[135] 'The Primordial Lithium Problem - A Big Problem for the Big Bang'
      Dr. Danny R. Faulkner AIG 01/15/ 2015
[136] 'Giant molecular clouds - A look at uniformitarian assumptions in star formation'
      John G. Hartnett PhD creation.com 03/15/2016

Preliminary Conclusion:

Thank you for your consideration. We hope you found herein some interesting perspective on Torah and Science. SPIRALL may be the strongest scientific explanation of the factual observations. The prevalent cosmological redshift (CR) of distant starlight and galactic rotation problem, absent missing dark energy & matter, fully align with:

- A physical universe 5,780 years old as of 5780AM.
- Starting by near a singularity or 'mustard seed' size.
- Hyper-dense Proto stellar and Galactic formation.
- CR & CMB are evidence of Cosmic Inflation Expansion.
- Proto-stellar & galactic formation preceding that CI.
- No ongoing CE subsequent to that CI – CE event.
- 'Magnetic Repulsion' may account for all the CI-CE.
- The universe has a center we are by that center..
- The universe is an approximate sphere (or spheroid).
- The 'visible universe' approximates the entire universe.
- We have the optimal view in the entire universe.
- 2B LY universe radius, % mid 1B min. & 4B max.
- Distant starlight departed the objects when closer. No more than 1LY distant per year elapsed post CI. When thousands, not billions, of LY distant.

Established by clear direct mass public revelation, an unbroken chain of transmission, the test of time, and intense scrutiny, the Torah narrative has not only withstood every challenge, but provides the best explanation of the natural observations. Consistency with Torah testimony is a good litmus test for other historic narratives and scientific hypothesis. See the Moshe Emes series to align Torah and Science in other fields.

May Hashem only increase grace, kindness & mercy, during this redemption process to peace and true knowledge.

Pearlman SPIRAL's **'Jiffy Pop'**
microwave popcorn, & the classic 'Leavening Raisin Dough',
analogies to help describe cosmic expansion (CE).[137]

While the dough that represents space expands, the raisins which represent the galaxies retain their relative positions in the expanding dough. The further the raisin - galaxy from any other, the faster they recede(d) from each other. For simplicity ignore orbital peculiarities.

If SPIRAL no ongoing cosmic expansion and no CE exposure of objects with SPIRAL light year radius I from any vantage point.

If SCM-LCDM the 'raisins' keep their approximate size, as CE distances them from each other. A hundred million years or so subsequent to a hyper-dense start, galaxies formed from accretion, the entire universe now at least 250 times the observable universe, now at a radius of 46.5B LY, over the course of 13B years. There is no preferred center. The raisin itself does not move, It is 'metric expansion' of space that causes the separation. Assume not 'gravitationally bound' to each other.

In SPIRAL like SCM is fine with the metric expansion of space having caused the distancing. A SPIRAL difference is we're by the approximate center of the entire universe, which approximates the approximate sphere that is the observable universe, and hyper-dense proto galactic formation prior to nearly all cosmic inflation expansion.

By the time the entre universe had a radius as little as the volume of a sphere with a radius of one light year rounded up (1k LY radius max). So the galaxies expanded along with the universe as a whole.[138]

So if SPIRAL not only a much cooler start, so no need to get bent out of (approximate spherical) shape. Based on the current given CMB temperature and vastly smaller volume of the universe to start.

---

[137] http://map.gsfc.nasa.gov/universe/bb_tests_exp.html
[138] See SPIRAL's 'MVP' hypothesis ours is the central and optimal view.

SPIRAL has a massive advantage with as many hyper-dense electro-magnetic repulsion zones as the number of galaxies or 'black-holes'. That are look back to long PAST hyper-density, rather than if SCM-LCDM from one hyper-dense start (that would likley get bent out of shape - 'flat' universe).

While light speed is limited to the speed of light, and is not subject to inertia, sound and matter may have been able to move apart from each other at speeds greater than light speed during CE-CI, due to inertia.

Proto-galactic hyper-density may have helped keep things together during the hyper-expansion.

To help visualize 'Special CE' use a group of magnets that repulse each other, where one by the center is stable due to offsetting forces.

SPIRAL's microwave popcorn 'Jiffy Pop' analogy:

Now think of the magnets as non-popped corn kernels, each representing a galaxy on a macro level. With the number of kernels = to the number of galaxies.

On a more micro-level each galaxy itself is comprised of smaller kernels corresponding to each stellar object therein... Imagine the kernels emit light.

If SPIRAL, the kernels expanded on both the macro galactic & 'micro' levels during Cosmic Inflation.

If 'Special Cosmic Expansion' the stellar objects emit light across the entire distance they came from. Just like in SCM, and SPIRAL's Preferred CE, where their light is subjected to CE, results in red-shift.

If Special CE they also emit light in the direction they moved, over the entire path they moved. Said light not being able to keep up, that was not reabsorbed into the object, should blue-shift.

Practical distinctions between Preferred & Special CE would be the view from distant viewpoints and CR distribution. See 'SPIRAL vs Hubble's Exhibit CE'.

'Special' or 'Preferred', the mechanism for CE could be electromagnetic radiation, magnetic repulsion.

If SCM we can no longer look back to where the CE started by the initial singularity as we are 'on the surface' and there is no center to the universe. So a speckled balloon that is not necessarily a sphere is a better analogy. The speckles 'galaxies' do not expand with the space fabric.

We are by the approximate center of the approximate sphere that is the 'visible universe' in both SCM and SPIRAL.

With SCM CE is assumed to be happening in every direction of the visible universe so the more distant galaxies are moving away from us faster than those closer by. Stars that were up to 4B LY distant 13.4B years ago (YA) are now 46.5B LY from us, the average OCE is at 42.5/4=3.27c (light speed). About an 11 fold expansion 42.5 B LY divided by 4B LY radius.

Think of a star 1 vs. 10 LY and double each. Now 1 vs 20. 1 vs 100.. Whenever we double a larger LY figure 'x' vs 1 LY we find the average expansion speed of the outer edge at radius x, as x times that of the expansion velocity by the radius starting at 1LY.

If we compare x to .5 LY it would result in a velocity of 2(x). If .1 LY it would be 10(x).. If we start 2LY to X the velocity will be .5(x). At 100LY vs x LY = .01x

Now compare the average recession velocity of two stars, as CE doubles their starting 13.4B vs 5k LY distance. The distant star will move 13.4B in the same time the nearby one moves 5k LY. If a uniform 'leavening of the dough' CE.

Per SCM a distant galaxy recession rate of 26,800 times that of the closest stars subject to CE, if starting 500k LY distance. The most distant CR light measured to date is from 13.4B LY distance shows 11.5 g (CR) on the scale where nearby stars subjected to CE show 0.5g.[139]

If SPIRAL: The entire universe approximates the visible universe we are by the center. The radius being up to about 2B LY. So the velocity differential due to CE between the closest and most distant CR'ed light is less than under SCM.

---

[139] http://arxiv.org/pdf/1603.00461v1.pdf Oesch, P. A. and team; 'A Remarkably Luminous Galaxy at z=11.1 Measured with Hubble Space Telescope Grism Spectroscopy'

The nearest stellar objects whose light visible here now was subjected to CE on departure is 5,778 LY to date. And increasing 1 LY per year.

If SPIRAL, without OCE assumptions, the entire universe stabilized post CI-CE at a radius of under 4B LY. We also use the raisin dough analogy on the macro level, the raisins represent the galaxies, which also expanded during the CI-CE epoch.

Our galaxy being by the center of the dough. We can also use it on a more micro level for individual galaxies. Where the raisins represent stellar objects, that also expanded (and rotated) during the CI-CE event, in each respective galaxy. Thus leaving the illusion of a black-hole by galactic centers.

Under SPIRAL we apply this raisin dough analogy with the difference the 'raisins' were much denser to start and the dough stopped rising (no ongoing CE) after the day four cosmic inflation expansion event.

Recall the Talmudic lesson each human life is equated to a universe. If each proto-galaxy started out the same human egg size we did, it's still larger than a singularity that contained the entire universe :)

If SPIRAL all the hyper-dense 'raisins' fit in a pre-CI-CE sphere with a radius of under one LY. Based on 5,778 age and continuous life supporting conditions lasting 7k years we're 82.5% to a goal for life on Earth.

The most distant stars are part of the observable universe, as the universe was made for us. So unless serving an alternate purpose, not much point in stellar objects beyond what could ever be detectable to us. So we can 'inquire to the ends of the universe'.

There is also evidence that the existing constellations were already visible to us by the earliest known 'Akkadian' civilization. No distortion over thousands of years of observation therein indicates there is no OCE.[140]

In SPIRAL the galaxies expanded during CE. This CE lasted up to 24 hours so is better described as a cosmic inflation (CI). CI being a form of CE.

---

[140] 'Torah Discovery Chronology' www.amazon.com/dp/B074Q6MJYF

The galactic rotation problem for SCM, absent the missing dark matter, stellar velocity and rotation attests to SPIRAL hyper-dense proto-galactic formation, thousands, not billions, of years ago, which does add up.

In SPIRAL proto-stellar formation preceded CE represented by this 'leavening of the dough'. The dough stopped rising after the day four cosmic inflation event.

All the CR light visible here & now, departed when the distant stellar objects, were the same LY distance from us, as the CE inflation epoch was years ago. Based on light speed at the standard speed of light.

If RCCF, the CE inflation event was 5,778 years ago to date, so 5,778 is the LY distance all the CR/CB light was from us when it departed when the 'raisins' were making their way to their stable orbital positions, up to the far edge of the visible universe, about 2B LY.

Change in density of the stars & intensity of their light factors should be considered in determining the age, formation & structure of the universe.

The condense proto-galaxies reached their mature size and placement about the same time, by the end of the CE, regardless if how far away they ended up. So the more distant a galaxy, the denser it was, and intense it's light, when it emitted what is reaching us now.

Density and intensity might have more of a role in regard to Olbers' paradox and light detection than with CR measurements, but just in case the theory or practice indicate otherwise, it may be something to be considered. The title of an interesting recent journal article on how density might affect CR.[141]

Under both SCM and SPIRAL calibrate for Peculiar Motions such as normal stellar orbits. The more distant the object the less relative effect so expect negligible unless nearby. Expect the same in the IU. If SPIRAL during CI much tighter and more recent thus OU still appears crisp.

---

[141] 'New Method to Assess the Luminosity Function of Galaxies'   by Dr. Jason Lisle

When determining the distance to the nearest stellar object whose light visible here & now was ever subjected to cosmic expansion:

Consider the change in density if SPIRAL.

Consider the spectrum between mid-Yellow normal light, including orange, and the start of the Red. So a lot closer than where the CR light is already 'Red'.

Exposure of light to water, or other sources of resistance, may have shortened it's wavelength so raised it's frequency. Resistance on light wavelength by or after departure, may have offset CR of distant starlight in our galaxy or nearby galaxies, in addition to, or instead of, 'gravitational bounding'.

With our narrative the proto-galaxies formed before water, the upper waters were in place before the Cosmic Inflation - Cosmic Expansion (CI-CE) event.

While the 'raisins' near and far expanded a similar amount on average to get to approximate their mature sizes. The CE of space meany what is more distant from the center expands faster, so the more distant galaxies moved away from us faster than those closer by.

That helps explain why if SPIRAL we might expect similar CR measurements to what SCM attributes to accelerating OCE. Is it because OCE is accelerating? Is linear? Is it because past CE, but not OCE?

Is it because we are by the center as per SPIRAL? or is SCM's there is no center and we are on the 'surface' and everything is moving apart?

Don't be surprised if starting day one the hyper-dense universe rotated once daily, as the Earth continued to do, once the universe of mature density post CI-CE.

As pointed out in HTP and Magnetic Repulsion: Each hyper-dense proto galaxy relates to the formation of it's electromagnetic field. That relates to the CI expansion epoch.

So, said fields approximate the number of galaxies. The CMB may be the residue evidence thereof. Smaller and more numerous fields are the better design and explanation.

Gravitationally Bound / Gravitational Binding
(+ see 'GRaB' on page 231)
May not be viable to explain lack of ongoing cosmic expansion in our galaxy region, or any other. [142]

Why should it be all or nothing if gravity was/is the reason? Where gravity is stronger than average in a gravitational bound place, there should be cosmic contraction that increases w/ gravity, and where gravity is weaker then average, cosmic expansion (CE) should start and increase as the gravity drops to zero, to the rate of hypothesized CE where there is no gravitational bonding and CE averaged a cumulative 3.26c if SCM. [143]

A larger problem for gravitational bonding under SCM-LCDM is the missing (non-existent?) dark energy is required/predicted to account for ongoing cosmic expansion outside gravitationally bound sectors. Now if SCM-LCDM the visible universe is now 46.5B LY radius but began within no more than 4B LY radius of us. SCM also holds galactic formation was via gravitational attraction accretion from an area as large or larger than the galaxy itself.

The number of galaxies is estimated to be between tens of billions up to a trillion or so. The distant ones have to be old or we would not see them. SCM overall assumes an even distribution of galaxies over large distances and most galactic formation was relatively early in history. Any recent ones formed may have been offset by deceased ones. So the problem is all those galaxies, or the matter they comprise of, was in an area at least 1,571 times as dense.

4x4x4=64. 46.5x46.5X46.5=100,545. 100,545/64 = 1,571

---

[142] 'Remaining Problems in Interpretation of the Cosmic Microwave Background' June 2015 Physics Research International and 'The influence of gravitational binding energy on cosmic expansion dynamics:..' June 2012 Astrophysics and Space Science 339(2) H. J. Fahr & Michael Sokaliwska

[143] "The influence of the cosmological expansion on local systems," http://arxiv.org/abs/astro-ph/9803097v1 Cooperstock, Faraoni, and Vollick

Two alternatives with a reasonable probability were: The cosmic energy was also 1571 times as dense and strong, in which case the expansion would have been vastly faster in the past. If the Galaxies could even form in that environment long ago there should have been a big rip!

A second reasonable alternative is the dark energy (assuming it even exists) was the same strength and density as it is now, and with the galaxies so much closer together, all/most of the universe may have been/stayed gravitational bound. More likely a lack of proportional dark energy offset would have long ago resulted in a big crunch!

Keep in mind a law of science aka the conservation of energy, means it was here, in some form, or it is not.

Failure to detect a spectrum, or even trace amounts, of Dark Energy and cosmic expansion within gravitational bound regions may mean one can not get from there (the young universe under SCM-LCDM) to here (the existing universe we actually observe today). Leaving SPIRAL as the more viable option between the two cosmological models.

SPIRAL does not interpret the factual observations as evidence of Ongoing Cosmic Expansion (OCE) so does not predict/require the missing dark energy. Thus the universe could have reached equilibrium after the end of cosmic inflation expansion between the gravitational and any other attraction and repulsion forces that would otherwise cause OCE or contraction of the fabric of space.

'Magnetic Repulsion', or the like, resulted in a cumulative expansion rate at speeds vastly greater than light speed, as the universe went from hyper-dense to mature size and density within a relatively short span. By day four end.

See postscript more on why the existence of dark energy and dark matter are internally inconsistent w/ certain empirical observations.

'CREST' Hypothesis

Cosmological Redshift (CR) in distant starlight depends on the cumulative amount of Cosmic Expansion (CE) it has been Subjected To.

While the validity of SCM should depend on this hypothesis being valid. If not SCM should not predict increased CR with increased distance. Under SPIRAL we would predict increased CR with distance whether of not CREST is valid.

The cumulative amount of CE is the product of the average rate of CE and the time span elapsed during which the CE occurred. If we know two of these three values we can get derive the third.

So light subjected to x LY distance of CE in one year should have a similar amount of CR as light from a galaxy subjected to one LY distance of CE for x years.

It may help to think of light being subjected to one LY distance of CE than a pause, than another and pause and so on. That way it is easier to envision each additional LY distance of CE as increasing the amount of CR.

Assume light departs a star 4B LY away and it reaches us after 13.4B years, and that star is now 46.5 B LY away. If Crest the light has CR based on it having been subjected to being subjected to 13.4-4 = 9.4B LY of CE.

The star moved away 46.5-4=42.5 B LY in 13.4B years, an average rate of 3.17 times the speed of light (c).

With a constant rate of linear expansion and a center, the greater the distance from center the faster the speed of recession and rate of CE.

The closer the distance the lower the rate of CE. So CR will be greater the greater the distance from center in a sphere as per SPIRAL even if a stable linear expansion. See the SPIRAL leavening raisin dough analogy.

If we are on the surface of a 'flat' universe as per SCM and a constant linear CE. If CREST is valid one would predict CR to increase with distance.

If CREST is not valid, linear CE and no center (as in SCM) you should predict uniform CR regardless of distance.

The CR does in fact increase with the distance of the source of the CR light. So CREST is valid, or SCM is invalid if a linear CE.

CREST hypothesis if valid, works regardless of if the CE is at a linear, accelerating, declining, or gyrating rate.

So a low steady velocity of CE over a long distance can result in a higher CR level than a higher velocity of CE over a shorter distance. Even a period of decelerating CE would continue to add to the CE until expansion stopped or reversed if ever any cosmic contraction.

So as long as we are looking at light that was subject to the same amount of cumulative CE it should not matter if the CE was at the first moment the light was departing, the last moment just before the light got here, or spread out in-between, the amount of CR should be similar.

With light speed not exceeding the speed of light under both SPIRAL and SCM, if never subjected to CE we should see light after 13B years light from stars that are now 13B LY away with no CR.

Now assume a Galaxy in place for over a year that is one LY away today, is subjected to 13B LY distance of CE all tomorrow. It should cause a 13B year delay until it's regular light can reach us again. The next 13B years it's light will have 13B LY worth of CE so be highly CR'ed.

If it moved away this fast there might not be much of a light trail unless it's light was intense enough. In SPIRAL the proto-stars and proto-galaxies were much denser. See our 'Proto-Stellar Blue-shift Offset Continuum' Hypothesis (CB).

So the light they omitted during a day four cosmic inflation event was much more intense than the light they omitted after expansion.

That is where the density of the proto-galaxies comes into play. If they were xB+/- times as dense as they departed, even if the departed rapidly and expand along the way, we should still get a much more visible light trail at first.

The first thousands of years, until a higher proportion of higher CR subjected to more CE light and stellar density decreased with distance during the CI event.

If CREST the amount of CR should be the same if the CE it was subject to happened over 1 day, 6B or 13B years.

By linear CE we mean a uniform expansion rate every distance out there will appear to an observer here to double in distance in the same amount of time.

At a fixed rate of CE the recession rate of distant objects increases. So acceleration away from the observer increases with distance. In a 'flat' universe with no center space near or far expands at the same rate of CE.

From the center in a spherical universe the further the distance the faster the rate of CE even with linear CE. If not clear please review the leavening raisin dough analogy.

By accelerating CE in the SCM assumption of a 'flat' universe with no center we mean the more distant the object the greater the increase in CE over and above the fixed CE rate of a linear expansion.

SPIRAL starting by the center in a universe that is a sphere even linear CE results in acceleration of CE with distance. The CE acceleration might have above and below linear CE and yet result in higher CR with greater distance.

Two issues SCM should have with accelerated CE are CMB blackbody temp depends on how much expansion an area was subjected to.

The more expansion the more it cools. Accelerating CE would leave nowhere close to blackbody temp uniformity with more distant CMB being much cooler than nearby readings.

A second issue is it would indicate we are the center of a sphere and falsify the SCM assumption premise over long enough distances the view from any point in the universe appears as the view from any other point.

SCM where there is no center to the entire universe:

If a fixed linear CE, distant CE is at the same rate as nearby CE. If no CREST distant light subjected to CE should have the same CR as nearby light subjected to CE.

If accelerating CE with no center, CR would decrease with departure source distance if no CREST, as this light departed longer ago when the CE rate would've been lower.

Yet the measurements clearly show the more distant light has more CR. So SCM needs CREST to be valid. The CR measurements falsify SCM if CREST is not valid.

SPIRL calibrates well to increased CR with distance. One variable would be how great or small the actual radius of the universe is once we remove deep-time assumptions. CR from starlight calibrated to measure 13.4 b LY years distant from stars formed 4B LY away now presumed to be 46B LY away might show the same exact CR as stars 4B LY or less under SPIRAL that formed with 1 LY away 6k YA.

Even with a reasonable amount of accelerating CE CR increases as the distance increases from a set point while the rate of increase as a % of total CR declines. Factual CR measurements can calibrate with SPIRAL if using SPIRAL instead of SCM assumptions.

If SPIRAL no matter how distant objects are now we are seeing their CR light from when closer. From a LY distance = to years elapsed from when the CE/CI ended.

SPIRAL **'CoMBO'** hypothesis for <u>C</u>osmic <u>M</u>icrowave <u>B</u>ackground Radiation <u>O</u>ptimal alignment.[144]

CMB near uniform blackbody temperature indicates past hyper-expansion of the universe. So aligns best with uniform cosmic inflation expansion (CI). Where something 1LY apart would take the same span to become 2 LY apart as something 100LY to become 200LY apart..

If CMB was the result of slower expansion, such as the alleged by SCM-LCDM ongoing cosmic expansion, it wouldn't have such a uniform distribution.

If CMB is from hyper-fast, but not cosmic, expansion, then there would have been a linear decrease in CMB temp. with distance. Still it would have near uniform CMB temp. from any fixed distance from us in any direction, be it 6k LY, or if SCM 13.x B LY. So pure super-fast distancing alone is not consistent w/ universal near blackbody CMB.

Another reason it had to be cosmic expansion (CE) type distancing, if speeds vastly greater than light speed 'c': Radiation, which includes visible light, is limited to 'c', and is NOT subject to inertia. Thus movement limited to 'c'.

CMB blackbody is thought to have an isotropic distribution throughout the universe. Highly redshifted up to non-detectable to the human eye. The redshift depends on the CMB temperature, which is a function of expansion,

If blackbody, on expansion, CMB cools. The greater the cumulative cosmic expansion subjected to (so a function of velocity & time) the greater the cosmological redshift (CR).

---

[144] Penzias, A.A.; Wilson, R.W. A Measurement of Excess Antenna Temperature .. Astrophysical Journal 142. 1965. I find interesting Arno Penzias, an observant Jew, was in on detecting the predicted CMB in 1965. Reminiscent of Hashem revealed important events via Joseph about 3,500, and Daniel about 2,400, years prior. CMB being evidence of Hyper cosmic expansion, being evidence of SPIRAL, a big part of the Moshe Emes series, Torah and science alignment.

Think of adding hot water to room temp water, blending to fill an (pretend perfectly insulated) thermos. The average temperature of the sum volume after, is a function of the volumes and temp of the hot and cold before. Now take that thermos and fill a thermos 2x the size, and so on.

Visible light & other radiation are limited to 'c'. So under both SPIRAL and SCM we'd predict visible CMB, here and now, from the distance = to when it was being subjected to cosmic inflation expansion, to exhibit CR.

Just like we predict/expect using SPIRAL distant starlight to exhibit an increase of CR with distance. As per SPIRAL the CR of the CMB and distant starlight was due to a CI cosmic expansion epoch.

The universe is thought to have been opaque, so cloud like, until it cooled down enough via expansion, so loss of density, to where it was no longer opaque at 'recombination'.

At 1/100,000,000 it's current size it's thought to have been as dense as air is at sea-level = 1.275 Kg/M3. So even more dense prior, the smaller the area, the denser, the denser the hotter, when a steady ratio of heat to volume. Regular water density = 1,000 Kg/M3. So at 1,000/1.275 = 784.3 denser, a density = water.

Under SPIRAL, as with SCM, the radius of the universe increased over a thousand-fold during CI.

If SCM it's assumed to have taken 400k years from the start of the initial singularity until the size our visible universe grew to by the recombination, about 1/1100? the current size.

Followed by stellar formation. Not a lot of time for stellar formation based on the low density, and SCM need the entire visible universe stellar formation, and light departure, was within 4B LY.

If SPIRAL the approximate mature modern density was in place by the end of day 4. The universe starting day one by a minuscule hyper-dense size. If opaque until 1/1100? the mature size, it coincided with Proto-galactic formation, starting as early as day one.

The universe expanded on day 4 from xLY, to 2B+/-LY, radius. With Cosmic Inflation (CI) expansion fast enough to account for CMB measurements.

The lack of CE (cosmic expansion) in our Galaxy may be a tell that ongoing CE (OCE) is an illusion. So we don't know if the lack of OCE in Gravitational bond galaxies, like ours, is due to gravity or CR being from past, not ongoing, CE as hypothesized by/in SPIRAL.

So real CE might overcome gravity, as it had CI energy/to expand. Real CE during CI with CI energy could account for the expansion of the hyper-dense proto-stars & galaxies hypothesized by SPIRAL into their approximate current sizes and orbits, by the end of CI.

While if SCM-LCDM the mechanism for CI and OCE is not clear. If SPIRAL we suspect electro-magnetic repulsion was the mechanism for the CI-CE. See SPIRAL's 'Magnetic Repulsion' hypothesis.

Dec. 2018 / Kislev 5779: The greater the distance claimed to the nearest departure point of visible light here/now ever subjected to CE the bigger the problem for SCM-LCDM.

A 46.5-4=42.5B LY radius of spreading, dissipation over 13B years of CMB. As the far end of the visible universe was 4B LY max distant 13B years ago. Yet a lot of the CMB was in gravitational bound quadrants up to xM LY dissipating at c not 3.26c.

So if SCM predict a checkerboard spectrum of greater CMB variation depending on size of any gravitational bound sections. If CMB not checkerboard, it favors SPIRAL over SCM.

If SCM 'by the edge' 1LY annual CMB dissipation, if not, no net dissipation as an even exchange of CMB from a like, but non-visible, sector. Assumes Copernican Principal. If an edge 13B YA it is still an edge, and was/is leaking CMB. Where not an edge, no leakage. So if SCM one should predict/require a more checker-board reading than found.

A testable prediction using CMB to see if there's no OCE (SPIRAL) vs yes OCE (SCM): If SPIRAL a stable radius to the most distant galaxies subsequent to the day four CI CE event.

That distance 2B LY +/- is much closer than the 46.5B LY most distance detectable objects if SCM, which assumes OCE.

If SPIRAL the radius of the spherical universe is expanding in regards to CMB at no more than 1 LY per year. As the CMB 'leaks' dissipating to beyond the most distant stars at 1 LY per year. If for some reason the CMB can't 'leak' into beyond space, CMB temperature, if blackbody uniform, stays steady.

If SCM, OCE averages 3.27 times the speed of light. +1 if annual leakage if by an edge 3.27+1 = 4.27 LY CMB dissipation.

If CMB temp. stays the same it means there is no leakage and no OCE. So SPIRAL not SCM.

Based on each models visible universe size, CMB temperature drops slightly less if SPIRAL. than if SCM.

If leakage and SCM the increase in area is 4.27LY radius expansion of a sphere 60BLY (46.5B+13.5B LY leakage).

If SPIRAL the CMB annual dissipation ratio is an increase of 1LY to the sphere up to a 2B LY radius.

While both tiny fractions, SCM sphere 4.27LY vs 1LY faster radius growth is offset by the area of the sphere with a radius 60B/2B max = 30 minimum times that of SPIRAL.

If SPIRAL to determine the actual size of the sphere that is the entire universe measure the annual temp change, solve for if 1LY leakage per year increase in the CMB radius.

Based on current CMB temp. initial temp. needn't have been as hot if SPIRAL as a far smaller universe and max thousands, not SCM billions, LY CMB leakage beyond.

If SCM or SPIRAL, and CMB is rising, it means CMB is not perfect black-body, or equipment/control/human error.

CMB near blackbody is widely accepted as evidence for a big bang model and a CI-CE event. You might think it is equally supportive of SPIRAL and SCM. However it may support SPIRAL & falsify SCM if the blackbody is evidence of no OCE.

The blackbody temperature depends on the amount of expansion. The more expansion the cooler the temp. Uniform expansion should mean a uniform temperature. Blackbody is widely taken as evidence for/of uniform temperature, due to a past hyper- rapid CI expansion epoch.

Yet even SCM agrees with SPIRAL there is no OCE in and near our galaxy. If SPIRAL there's no OCE period. SCM holds no OCE in any galaxy. So speculates gravitational attraction negates OCE. If so, why is there no evidence of OCE in-between the local Galaxies that we can observe the best? Perhaps because there is no OCE!

Now if areas subjected to gravity like galaxies have not had OCE, but elsewhere there has been OCE, as the SCM explanation of CR need hold, then the CMB was cooled by expansion in many areas many times more than in many other areas. So if SCM were true we would predict/ expect no even CMB blackbody temperature.

Other examples where we could predict temperature differences: On a hot sunny day in places where some is in tree shade. In the summer where parts of the region are by a cool ocean. Between a well insulated house and the exterior when the temperature is dropping outside..

In summary SCM depends on SPIRAL's CREST hypothesis being valid and CMB blackbody being invalid. If CREST is not valid: either the SCM premise the universe has no center is falsified or CR does not increase with distance, which contradicts the empirical observations. So SCM between a rock and a hard place.

That CR increases with distance aligns with the universe having a center and our Earth-sun ecliptic being by it. If CREST is valid the CR under SPIRAL approximates the CR under SCM after calibration for the assumptions of each model.

CMB near blackbody being valid aligns better with SPIRAL where there is no OCE subsequent to the end of a cosmic inflation event. It would falsify SCM which holds no OCE in some areas such as galaxies but not other areas. CMB approximating blackbody distribution favors SPIRAL.

With CREST we saw either CREST is valid and SPIRAL CR approximates CR under SCM. For example the most distant CR light we are seeing now would have the same amount of CR under SPIRAL at a radius of 9.4 B LY as SCM from 13.4 B LY .

If CREST is not valid than SCM is invalid as it does not align with CR increasing with distance. Thus leaving SPIRAL as the only viable alternative between the two that aligns with increasing CR with distance.

So we simply calibrate the CR based on the SPIRAL model instead of the falsified SCM model. SCM only aligns with a steady amount of CR regardless of distance, as explained in CREST, if CREST is not valid.

See postscript for more on why the existence of dark energy and dark matter are internally inconsistent w/ certain empirical observations.

- CMB fits best with a big bang start from a small area size.
- CMB fits best with a CI hyper CE expansion event.
- CMB fits best with a constant linear CE, be it OCE or not.
- CMB good if the CE was 1k times less than if SCM.
- CMB good if dissipated 1k times less than if SCM.
- CMB more likely to indicate the universe is a sphere with no OCE, a smaller, more closed system, than under SCM.
- CMB more likely to indicate no OCE as any change anywhere in OCE over time/space will leave CMB erratic.
- CMB favors YeC if gravity over time congregates CMB
- CMB favors YeC as more time = less chance linear CE.
- CMB more likely with SPI of SPIRAL as stellar formation if SCM with gravity would leave CMB erratic.
- CMB CREST hypothesis: For testable predictions of the rate of CMB temperature change in relation universe size.
- CMB detectable here now if SCM/OCE departed 4B LY away, 13B years ago and traveled 13B LY to get here.
- CMB detectable here now if SPIRAL is from LY distance = to years elapsed subsequent to CI, as no OCE.
- SPIRAL caps visible CMB departure point LY distance to nearest object whose starlight was influenced by CI. (CR..)
- If RCCF CMB departure point 5777 LY/years to date.

See note at the end of description of Exhibit D on CMB.
See SPIRAL 'CoMBO' that concludes this section.

CMB 'Axis of ~~Evil~~ Justice'
(+ see recent SPIRAL articles on Radio and Quasar Dipoles)
June 4, 2019 / 1 Sivan, 5779

Gen. 1:3 Primordial light of creation[145]
K' Shlomo in Mishlei 13:9: 'The Light (teaching..) of the
righteous will rejoice..'[146]
'Rebi Yosi said, "Woe to the created beings who see but do not
know what they see, who stand but do not understand upon what
they stand . . ." (Chagigah 12b) ' ..

Just days after the March 20, 2013 Planck results release,
SPIRAL cosmological redshift hypothesis and model crystallized.
Empirical CMB observations confirming Earth as the universe's
focal point, was a helpful piece of the puzzle.

While it is true both SCM-LCDM and SPIRAL align with
the cosmic inflation that CMB attests to, it is the spherical
Geocentric universe SPIRAL that the details of that CMB
corroborate, and not the flat on the surface, Copernican Principle
without center assumption based, SCM-LCDM.

'With ten utterances (equations, as aleph-beis alpha
numeric) Hashem created and organized the physical universe
over the course of week one.'

'The one-in-100,000 variations from sound waves acoustic
signals observed in the CMB are of exactly the right amplitude to
form the large-scale structures we see today' (condensed).[147]

---

[145] Mishlei (Proverbs) 13:9 and T' Chagiga 12b see above
Isaiah 5:20-21 '.. woe to those who call good evil and evil good..'
especially if the intent is to obfuscate the truth, which obstructs the vulnerable
from observance of the eternal Torah covenant of 7-613.

[146] www.chabad.org/library/bible_cdo/aid/16384/jewish/Chapter-13.htm

[147] http://background.uchicago.edu/~whu/SciAm/sym2.html

1965 Penzias and Wilson announce the measurement of the CMB.
1980 Guth Inflation hypothesis re Horizon Problem conundrum.
As allows all points to be in causal contact with one another
initially..'[148]
2005 'Axis of Evil' dubbed:
'Some anomalies in the background radiation have been reported
which are aligned with the plane of the Solar System, which
contradicts the Copernican principle by suggesting that the Solar
System's alignment is special.[6] Land and Magueijo in 2005
dubbed this alignment the "axis of evil" owing to the implications
for current models of the cosmos, ..

We are unable to blame these effects on foreground
contamination or large-scale systematic errors. We show how this
reappraisal may be crucial in identifying the theoretical model
behind the anomaly.[149]
2013 Geo-centric universe CMB alignment with Earth-sun
ecliptic plane quadrupole and Equinox plane monopole, verified.

Using the current consensus CMB temperature and avg.
current density of normal matter in the universe as givens:

SPIRAL at 2B LY radius of the entire universe is 2.5M
times smaller than SCM visible one at 46.5 LY radius area (200).

That means whatever temperature the competing
hypothesis SCM reached at any particular net area size, during
cosmic inflation /cosmic expansion, SPIRAL reached that
Temperature 2.5M times earlier ie when 2.5M times smaller.

[148] 'Cosmic Anisotropy: A History of the Axis of Evil' researchgate.net
     publication/267097146 Damian Sowinski March 2013
[149] 'The Axis of Evil' Kate Land, Joao Magueijo Feb 2005.
     Published in Phys.Rev.Lett. 95 (2005) 071301 astro-ph/0502237

Planck 2013 results:

'The two fundamental assumptions of the standard cosmological model - that the initial fluctuations are statistically isotropic and Gaussian - are rigorously tested using maps of the cosmic microwave background (CMB) anisotropy from the Planck satellite...

Although these analyses represent a step forward in building an understanding of the anomalies, a satisfactory explanation based on physically motivated models is still lacking.'[150]

MIT Professor Max Tegmark concludes these empirical CMB observations show we occupy a 'very significant' position in the universe.[151]

Combine that with the YeC Moshe Emes framework for Torah and Science alignment showing our life spans are very significant in relation to the age of the universe.

So how much more than 'very significant'. Our role being the very reason the universe exists.[152]

With but 200 rounded generations of man have spanned the history of the entire universe. Adam created in full stature on day six being the start of 1 anno-mundi and this year 5779.

---

[150] https://arxiv.org/abs/1303.5083 Isotropy and statistics of the CMB Planck Collaboration Mar 20, 2013.

[151] 'The Cosmic Microwave Background proves intelligent design' (Enoch'sMuse). https://youtu.be/y2AwSIbtv38 via @YouTube
Professor Max Tegmark comments at the 6:12 - 8:12 minute mark.
Gen. I:4-5 could the division of light and darkness relate in any way to CMB Warm & Cold zones neatly divided along the Ecliptic plane extending to the far reaches of the universe? 8:12 – 10:00 min. mark

[152] T' Berachos 6b, R' Elazar on Ecc. (Koheles) 12:13 '..for this is all of man' The entire universe was/is on behalf of the righteous person..

26 generations from Adam (1-930) to Moses (2368-2488). 5779-2368 = 3411 / 20 = 171 + 26 = 197, so 200 rounded generations to date. Assumes 20 years per generation, after Moses born.

CMB radiation show both expected and unexpected anisotropies in the CMB.. aligned with features of the microwave sky,.. correlated with the plane of the earth around the sun. ..– the plane of the earth around the sun – the ecliptic. That would say we are truly the center of the universe.' .. '

Data from the Planck Telescope published in 2013 has since found stronger evidence for the anisotropy.[17] "For a long time, part of the community was hoping that this would go away, but it hasn't," says Dominik Schwarz of the University of Bielefeld in Germany.[18] '..[153]

Good for SPIRAL 'MVP' hypothesis vs Copernican Principle. Keep in mind Copernicus himself held by an Solar centric universe which is much closer to SPIRAL Earth-Sun ecliptic universe than to one without a center. So deceptive to invoke his name, for that disputed principal assumption.

'The detailed particle physics mechanism responsible for inflation is unknown.' "The modern explanation for the metric expansion of space was proposed by physicist Alan Guth in 1979, ... all the observations that strongly suggest a metric expansion of space has occurred,'..[154]

[153] https://en.wikipedia.org/wiki/Axis_of_evil_(cosmology)
[154] https://en.wikipedia.org/wiki/Inflation_(cosmology)

'CMB is .. evidence of (a).. Big Bang origin of the universe. When the universe ... was denser, much hotter... Although many different processes might produce the general form of a black body spectrum, no model other than the Big Bang has yet explained the fluctuations. As a result, most cosmologists consider the Big Bang model of the universe to be the best explanation for the CMB. .. '[155]

That SCM-LCDM alone aligns with CMB is false, unless you include SPIRAL in your definition of 'the Big Bang' which I doubt they intended/considered. As it aligns as well, or better as we find, with SPIRAL.

Surface of the last scattering: We detect the CMB here that has traveled the LY distance = to years elapsed from the CMB photon distribution, early in history. If SPIRAL from SPIRAL LY radius 'I'. Predict as of (June 4, 2019 / 1 Sivan, 5779) from 5,779 years ago and from a like 5779 LY distance. With CMB now spread over a distance of 1B rounded LY, radius that the universe attained when it attained mature size and density (gravitational bound equilibrium) by the end of day 4.

If SCM-LCDM from x LY, 13.y B years ago and traveling the same 13.y LY distance (due to their ongoing cosmic expansion assumption) CMB now spread over a radius of 46.5B LY due to the same ongoing cosmic expansion assumed.

So, if SPIRAL, a much smaller universe now means it never was nearly as hot as SCM-LCDM when at any equal size. 'CMB redshifted photons indicate that the universe is expanding. (SPIRAL – expanded) This expansion implies the universe was smaller, denser and hotter in the distant past.'[156]

---

[155] https://en.wikipedia.org/wiki/Cosmic_microwave_background
[156] https://wmap.gsfc.nasa.gov/universe/bb_tests_cmb.html

Consensus SCM assumes dark matter and energy.

A 20:1 ratio if they exist vs if they do not, as if SCM-LCDM the invisible dark matter and energy comprise about 95% of matter/energy. So if SCM prior to recombination, when opaque, the opaque matter/energy included both what would become dark matter and energy and regular 'visible' matter and energy.

So if dark matter and energy do not exist, the density and temperature of the opaque phase of the universe was 20 times less dense and cooler per fixed area.

So based on this one variable, a like sized stage of the universe was 20 times as hot under SCM-LCDM than if SPIRAL. Which is the same as saying at the same temp. the size if SCM would be 20 times that if SPIRAL.

Under what conditions does it matter if the visible universe approximates the entire universe vs just 1% +/- thereof?

As if we start with a sphere prior to cosmic inflation expansion and stay a sphere (SPIRAL) vs 'Flat on the surface SCM-LCDM with total area well over 100x (125Mx?) the visible universe will greatly affect the change in temp due to the difference in size vs density post expansion..[157]

IF SCM-LCDM and 'flat' on the surface the entire universe must be at least 10x and 1,000x the size within reason the size. Test for all values, here assume 100x ..

So when the entire universe went from a tiny hyper-dense size to a total area of 1 LY radius based on this one variable alone it was Xx as hot if SCM-LCDM than if SPIRAL where the entire universe approximates the visible universe.

---

[157] If SCM 13B LY travel distance and year lapse of visible CMB here and now, angular measure between CMB hot spots means entire universe 500x the radius so 125M the volume, as the visible universe. Don Lincoln PhD

So 20X and 100x = 2,000x as hot if SCM-LCDM if a like sized opaque area. If the same temp than the opaque area was 2,000 x larger if SCM than if SPIRAL after factoring in the entire universe and if the disputed dark energy and matter (only predicted / required if SCM, not if SPIRAL) exist.

My understanding of the current consensus CMB is recombination when the universe stopped being opaque was when it was 1/1100 it's current size.

If SCM, assume a 46.5B LY visible universe radius. 46.5B /1100 = 42M LY radius at recombination

If SPIRAL assume a 1B LY radius, but test for values of 4BLY maximum, and 40M+/- LY minimum, radius.

Assume they mean 1/1100 of the volume/area (not radius) of the sphere. If so each doubling of the radius = an 8 fold increase in volume/area. From 32B LY to 46.5B is a 3 fold increase.

2B to 46.5B radius is 2>4>8>16>32>46.5 doubles 4.x times. 8,64,512,4,096,32,708.. So the sphere w/ a Radius of 46.5 B LY has 12,288 that of one 2B LY radius and 98,304 that of a 1B LY radius and 786,432 that of 500M LY radius..

Assumes all energy/matter opaque prior to recombination,, but what if some proto-galactic formation, prior to or during the Opaque phase and prior to day four cosmic inflation expansion.

That should mean a lot of the potential heat rather than straight heat was already in the proto-galaxies so even cooler than the 2000x cooler under SPIRAL vs SCM based on factoring in the entire universe and no reason to predict/require any of the missing dark energy and matter.

If SPIRAL no ongoing cosmic expansion, we are seeing CMB from 5779 years ago when it was 5779 LY distant. The CMB was near black-body evenly spread 5779 years ago, to date, over the sphere of the visible universe. That approximates the entire universe, we are at the approx. center of, that has a radius of 1B LY +/-.

So if there is an Opaque edge to the visible universe, beyond which nothing is, or will be, visible, that would be at a distance of 2B+/- LY. 5779 LY as of 5779 anno-mundi,

increase x2	increase x8	divides by 2	divides by 2
1	1	1	46.5B	2B
2	2	8	23.25B	1B
3	4	64	11.6B	500M
4	8	512	5.8B	250M
5	16	4096	2.9B	125M
6	32	32708	1.45B	63M
7	64	261664	727M	31M
8	128	2093312	363M	16M
9	256	16746496	182M	8M
10	512	133971965	91M	4M
		times less	SCM	SPIRAL
		volume	Radius	Radius

So, at 1/100M the estimated current size where the density was alleged about that of air at sea-level.

SCM: CMB density of air when about 100M radius.

CMB density of water when about 10M Radius.

Air being 0.1275% the density of water. 1.275/1000

So radius has to half a bit over 3x to get over 1/512.

SPIRAL: 1/100M 2B radius volume about @5M radius. If CMB dense as air @5M, then water density @500k radius.

Now factor in no dark matter and no dark energy required or predicted if SPIRAL. In SCM they comprise 95% of matter so reduce size by a factor of 100/.05=20.

Now factor in the entire universe approximates the visible universe if SPIRAL, not 1/10 to 1/1000 +/- so reduce size by the average 10/1000=100 by a factor of 100.

When we think about it why would Hashem not start creating the proto galaxies in an opaque hot dense phase?

Think of an iron forge/foundry. The air and oxygen distribution is evenly distributed enough for the workers to breathe, not to far away from the hot metal being processed, but we may use a bellow and adjust the heat...

Fish may lay their eggs and flutter over them, where the water overall is evenly distributed.

Morning fog may be a way to visualize opaque...
If the warmth of the sun speeds up the molecules, so they expand, so lifts/dissipates the fog.

Water being a better conductor than air. So why wouldn't CMB at the density of water be a better conductor for proto-galactic formation than when at air density? Far more so than when more dissipated after recombination.

Dense enough, and hot enough, for millions of years worth under SCM, to be finished in hours.

So why wouldn't Hashem create and 'Merachpeset over the water' density opaque universe, forming the proto-galaxies?
158

This reminds me of the saying 'fight fire with fire'. As well as Torah on Hashem protecting Abraham in the furnace at Ur-Kasdim and later Daniel's friends in Bavel.[159]

So in the hyper-dense, hot formative stage, resonating hot on hot could cool, create heavy elements, exchange heat for heat potential,..

A hotter/denser opaque phase to the universe that quickly cleared up by or during the day four cosmic inflation expansion by the end of which mature size/density reached.

While the proto-stellar objects during an opaque phase wouldn't be visible through a human eye, the rotation and positions would be visible to Hashem. Think how whales can hear/detect/locate each other from well beyond eye-sight.

---

[158] Gen. 1:2 maybe 'water' here references the density of the primordial matter & energy, which included all proto-water.
[159] Gen. 11:28, Nehemiah 9:7, Gen. Rabbah 38:13, Daniel 3:8-33

Conclusion the opaque phase, CMB density and temperature:

Assume opaque starting day one, during proto-stellar formation, and ended by/before midday four.

SPIRAL's 'HTP' hypothesis: heat time & pressure, 3 variables in stellar formation, fits well with proto-galactic formation during a hyper-dense opaque phase.

If the Opaque phase ended when the universe 1/1100 the current size, perhaps too late after for stellar formation if SCM, where even the most distant visible objects omitting light did so within 4BLY from us. As just 1/4096 the size @ 3BLY radius based on SCM's 46.5B LY radius.

An opaque phase provides another explanation, aside from hyper-density and gravity, how/why the proto-sun could have started prior to day four, yet not visible yet.[160]

Opaque no impediment to proto-stellar observation by Hashem by day one. The opaque veil lifted for us day four.[161]

If someone in the N. Arctic extended daylight can know days/nights elsewhere and of the rotation of the Earth in relation to the constellations, so could Hashem day one.

---

[160] Gen. 1:14-18 By end day 4, visible to a human observer on Earth.
Isaiah 40:12 '..meted out the heavens with a span' is additional corroboration of SPIRAL's many scientific, scriptural and Torah Mesorah tradition testimony arguments of the universe approximating mature, modern, density and size, by end day 4.

[161] Psalm 139:12 'Darkness obscures you not, night, light as day to Hashem..'

Once Hashem has determined something, it is as a done deal. Thus we find the placement of the sun and moon starting based on year of Tohu - conception. So how much more so after the creation of time and space on day one.. is light /dark, day/night, considered as starting day one, even if a human could not observe till later in the week.[162]

Even once we could observe it our perspective is limited. Yet Hashem lets us set the time / calendar...[163]

Could the division of light and darkness relate in any way to CMB Warm & Cold zones neatly divided along our Ecliptic plane that extend to the far reaches of the universe?[164]

CMB alignment with the Earth-Sun ecliptic is very strong corroboration of SPIRAL's 'MVP' hypothesis, the Earth-Sun ecliptic is the approximate center of the entire universe, not just the visible universe.

The current SCM and SPIRAL agree, any opaque phase, and Cosmic Inflation expansion, were relatively early in the formation of the universe.

---

[162] Jeremiah 1:5 'At the womb I knew you'. Gen. 1: 2-5 + 14-19
   'Understanding The Jewish Calendar' Rabbi Natan Bushwick
   'The Jewish Calendar' Rabbi David Feinstein Sh'lita
   'Bircas HaChammah' Rabbi J. David Bleich Ph.D
[163] Leviticus 23:2 & 4 which we are to declare / set.
[164] Gen. 1:4,5 day one.

SPIRAL 'CoMBO' on CMB (start page 119)
(**C**osmic **M**icrowave **B**ackground **O**ptimal alignment).
Comparing SCM-LCDM vs the competing SPIRAL cosmological model on recent findings that cap primordial heat and density.[165]

Assume CMB attests to a hyper cosmic expansion 'inflation' epoch relatively early in the history of the universe.

Start from when the two models were at the same exact small volume (NOT the same density).

If SPIRAL NO dark energy and matter are predicted or required. So the universe is 5% (= 1/20) as dense as if SCM. So SCM is 20 times less parsimonious.

Based on CMB temp. the volume of the entire universe if SPIRAL at a radius of 1B LY then SCM is 1370 times as large as it's 46.5B LY visible universe radius, is (rounded) 100M (1370) times the volume of SPIRAL.[166]

CMB temperature and view is based on the current, not past, size/volume of the entire universe.

The size and density parsimony advantage if SPIRAL vs SCM-LCDM is (20) x 100M x 1370 = 2.74 Trillion : 1.

After taking into account SPIRAL's straight line entropy advantage of 13.8B/ 5,782 = 2.38M:1
2.74T x 2.38M = 6.54E+20.

An 6.54 Quintilian : 1 massive parsimony advantage of SPIRAL over SCM-LCDM. Not including the many other scientific advantages of SPIRAL over SCM.

---

[165] New Inflation Data: Trouble for Cosmology?
   https://youtu.be/IIT5Veqe3Xg Dr. Brian Keating Nov. 12, 2021
[166] Based on angular measures between CMB hot-spots (assumes SCM distance assumptions). Fermilab Don Lincoln PhD Dec. 22, 2021
   Secrets of the Cosmic Microwave Background https://youtu.be/ri2LIEjXhmE
   See SPIRAL 'SAFETY' on flatness.

Cosmic Microwave Background Radiation (CMB), near black-body (isotropic) distribution is due to a hyper-cosmic expansion epoch early in the history of the universe,. The current temperature is a given. We detect these red-shifted photons here and now from all directions, that have traveled the LY distance = to the subsequent number of years elapsed. Every-time the area of the entire universe doubled, the CMB radiation temperature halved. Working backwards the temp. doubles with each halving in size of the universe. Here we see size & dating where SPIRAL & SCM equate.

SPIRAL: The visible, approximates the entire, universe.
A 1B+/- LY (Light Year) radius of the entire universe (0.5B rounded minimum and 4B maximum).
The universe attained mature size & density by day end 4.
CMB photons were emitted before the end of day 4, and as early as mid-day 1. Electro-magnetic radiation early day 1.

SCM-LCDM (Standard Cosmological Model): 46.5B LY the radius of the visible universe, to date.
The entire universe is at least 250 times the area of the visible universe.[167]
CMB photons were emitted after about 375k years.[168]

---

[167] www.technologyreview.com/2011/02/01/197279/cosmos-at-least-250x-bigger-than-visible-universe-say-cosmologists/
[168] CMB emitted relatively early (375k yrs. if SCM) post Big Bang.
https://wmap.gsfc.nasa.gov/universe/rel_firstobjs.htm

SCM   46.5B LY radius x 250 is 25M times the area of SPIRAL radius of 1B LY.
CMB temp. if SPIRAL at 0.9B LY radius after 4 days, equates to after 375k years if SCM.
CMB temp. if SPIRAL at 0.567B LY radius after 1 days, equates to after 375k years if SCM.
SCM the entire, at 342.5 times visible, universe CMB temp. equates to SPIRAL after 4 days.
SCM the entire, at 1,370 times visible, universe CMB temp. equates to SPIRAL after 1 day.
CMB emitted after 274k years if SCM (at 250x visible) equates to after 4 days if SPIRAL.
CMB emitted after 68,493 years if SCM (at 250x visible) equates to after 1 day if SPIRAL.
Math (figures rounded):
         46.5x46.5x46.5=100k(B)(250)=25M(B)/1B=25M.
         375,000x365=136,875,000.
136.875M/25M = 5.475   4/5.475=0.73 .9x.9x.9=0.729
1/5.475=0.1826   .567 cubed =.1823
136.875M/4=34.218M   34.218M/25M=1.37
250x1.37=342.5   342.5 x 4= 1,370.
25M/365=68,493   68,493x4 = 274k.
         In any 1 of these 3 cases (or adjusted combination thereof): SPIRAL radius x, or SCM y times larger than the visible universe, or CMB emitted after z years, the temperature of the universe at the time of the emission of CMB photons if SCM, equates to if SPIRAL.

End:
SPIRAL vs SCM on the timing of the emission of CMB photons.

CMB mapping aligns best with, and attests to, SPIRAL
Dated Feb. 06, 2023 / 15 Shevat, 5783
Feb. 6, 2023 / Tu B'Shevat 5783 updates:

'"features" imprinted in this surface of last scatter (i.e.-regions that were brighter or dimmer than average) they will remain imprinted to this day because emitted light travels across the universe largely unimpeded'.[169]

Both models deal with the visible universe & a hyper-dense start:

If SPIRAL: CMB departure from SPIRAL light year (LY) radius 'i', i yrs. ago and 1B LY of cosmic expansion subjugation. CMB photons that departed toward us w/in 'I' LY already passed us. Those beyond radius I passed radius I 'I' years ago. As did all the other visible objects beyond radius I. Overall, the more distant now, the earlier on cosmic inflation day 4 it passed radius 'I'. We can/do map CMB as it reflects 99.99% of the visible universe. The portion of a 1B LY radius sphere less a 6k LY radius portion.[170]

If SCM-LCDM: under 00.01% CMB visibility, departure from outer edge, 13.7B LY of travel distance, starting 13.7B years ago, and (2023-1965 = 58 yrs.) 58 LY / 46.5B LY of data (after subjugation to cosmic expansion that averaged over 3LY per annum, as now assumed at a radius of 46.5B LY). CMB would only map a very narrow shell of the visible universe. CMB still in transit here, began beyond the edge of the visible universe.

---

[169] https://map.gsfc.nasa.gov/media/990053/index.html

[170] T' Chagiga 12b may give additional meaning to the 'daily renewal of creation' in context of Isaiah 40:22. Day 4 on display daily!
SPIRAL finds the universe attained mature size and density 4/365.25 (SPIRAL LY radius 'I') a fraction into history at a radius of about 1B LY, 'I' years ago. I = nearest LY departure point of any visible light ever subjected to any cosmic expansion. Predict 'I' = 6k rounded.

In the classic leavening raisin dough analogy, the metric of space expands as galaxies stay in their relative positions. CMB has dissipated as much here as at the edge of the visible universe over the same time.

Conclude: SPIRAL aligns best with CMB mapping.

Galaxy sized voids in CMB may attest to SPIRAL = hyper-dense proto-galactic formation PRIOR to cosmic inflation. If the voids are 'flat' it doesn't mean the universe is. Perhaps primordial association with a 'flat' disc galaxy?

CMB near black-body (isotropic) distribution attests to a hyper-dense start followed by hyper cosmic 'inflation' expansion.

SPIRAL is the best model to determine the start size and temperature of the physical universe. The duration of the hyper-cosmic expansion epoch. The mature size and density, at the end of that epoch with universe at near the current temp.

So CMB helps establish SPIRAL cosmological redshift hypothesis and model as the new standard in cosmology to beat. SPIRAL crushes the existing consensus champion competing hypothesis SCM-LCDM. Based on probability explanation of the empirical cosmological observations and internal consistency.

Torah Links, Etymology and Insight:

Genesis 1:2: Darkness on the TaHoM as hyper-dense, so light trapped. Alt. = Opaque phase of CMB, dense as water?, MiRaCHaPheS fashioning/incubating the hyper-dense proto-stars/galaxies.. as a bird (Rashi) or fish, hovers over it's nest.

Gen. 1:14-19 'KochaV' heavenly luminary/star. Etymology: dim? channel/line? pushed? bent? so stellar light trails?

'MeOreRote' may also refer to the light-trails resulting from CI cosmic inflation expansion 'stretching' of the heavens, as much as to the luminaries, the light trails emanated from.

Think SPI-RAL, thus the Kochav (visible light-trail) is in the 'Rakeah', even if the stellar object's beyond it, or spent. Implicit is the light from the most distant luminaries reached here, & a mature size &density universe, by day four's end.

See SPIRAL on parallax, the sun's gravity bends the light trails into a 'tent'! Also no ongoing expansion in a 'tent'.

Gen. 17:1 revealed name Sheamar 'Di' to end CE, day end 4.

Numbers 24:17 'KoChaV MeYaacov' (star of Jacob)

Deut. 4:32 'Inquire to the ends..' implies a center & no OCE..

Deut. 32:9,, CheVeL - measuring cord

Joshua 17:14 CheVeL – measuring cord

Judges 5:20 '..stars in their 'MeSuLot' (orbits?)..

Samuel I 17:5 KoVaH – helmet/cover

Psalm 8:4 'Kvunout'? Places/lines/trails/directions/orbits?

        Moon & Stellar objects You've set in place (No OCE).

Psalms 9:9, 96:10, 98:9 'b'meiShaRim'im' measured ways?

Psalm 19:2 'The heavens declare Hashem's glory & handiwork.

Psalm 19:3 '..liiLa YachVeh-Daas' night a bond to knowledge.

Psalm 19:4 Encoded therein as reveled at Sinai. See Hirsch

Psalm 19:5 '..KaVam..Melehem..' their light trails cover the Earth

Psalm 19:5 Hashem put the sun to bend their light-trails as a tent.

Psalm 19:5 No dark side of the Earth (unlike the moon's).

Psalm 19-5 Earth ecliptic 'tent' a reference the Earth Sun ecliptic.

Psalm 19:6 'sprinting' may references the Solar procession.

Psalms50:6'..heavens declare H' righteous & Judge & 'Rock'

Psalms104:2'covering w/light..stretched the heavens like a curtain'

Psalm 115:164 MiSuLah – pathway/highway
Psalms147:4-5 H' counts/names the stars w/infinite understanding
Psalms 148:6 VaYaMiDame stood/stopped the cosmic expansion
I Kings 8:27 ShamaYim & Shmei Shamayim outer, outer space ..
Isaiah 28:10 ..KaV L'Kav, Kav L'KaV..
Isaiah 34:4 'vNiglu' revealed. Visible, = entire, universe..
Isaiah 40:3.DeReCh(road) ..YiShRu(straight) MiSilah(path)
Isaiah 40:12 'meted out the heavens with the span' (now set)
Isaiah 40:22 'Spreads the Heavens like a fine curtain (CR/CI
        lower frequency/longer wavelength?)..tent (to static?)..'
Isaiah 40:26 Barah(created), Motsei(brings forth), MiSPaR-SVoM
(number -legions), (stars/galaxies visible light trails?) each by
name, none missing..
Isaiah 44:24-25 Noteh (spread out, via CI?) the heavens.., &
        confounds wiseguys. Isaiah 5:12 & T' Shabbos 75a
Isaiah 45:12 NaTue SH-MaYiM (CI)..TsaVitsi (command HTP..)
Isaiah 48:13 'meted out (now static) heavens w/ right hand-span..'
Isaiah 57:14 SaLu-SaLu pave a road - pave a road
Isaiah 66:1 Hashem:'Heavens my throne, Earth my footstool'
Amos 9:6 'Who built His upper-stories' Ma-sL-Loso
Zechariah 14:7 one day, than perceived as light
Jeremiah 10:12, 51:15 'Noteh' stretched out the heavens.
Job 9:7-9 uncovers deep things from darkness.'Noteh Shmym'
Job (Iyov) 12:22-25 they grope in dark.
Job 26:11 rebuke-enough (T' Chagiga 12a) Job:
37:7 'morning-stars' Rashi: 'Mthila (at start) KoCvei Ohr (stars)'
Evening prayers before Shema: '..miSaDeR es haKochaVim
'bimiShMoRosaehm'.. (arranges the stars in their watches).

T' Chagiga 12a:
Rav Yehudah 'expanding continuously from.. till'.
Heavens & Earth made simultaneously. Universe start spool size..
Leviticus.8:3 Midrash Tanchumah P'Tsav 12: Start eye pupil size..
Gen.1:3 RambaN: Entire universe start like a very small point..

Talmud Chagiga 12a
'.. But was the light created on the first day? For, behold, it is
written: And God set them in the firmament of the heaven, 26 and
it is [further] written: And there was evening and there was
morning a fourth day 27 — This is [to be explained] according to
R. Eleazar. For R. Eleazar said: The light which the Holy One,
blessed be He, created on the first day, one could see thereby from
one end of the world to the other; but as soon as the Holy One,
blessed be He, beheld the generation of the Flood and the
generation of the Dispersion, 28 and saw that their actions were
corrupt, He arose and hid it from them, for it is said: But from the
wicked their light is withholden. 29 And for whom did he reserve
it? For the righteous in the time to come, 30 for it is said: And
God saw the light, that it was good; 31 and 'good' means only the
righteous, for it is said: Say ye of the righteous that he is good. 32
As soon as He saw the light that He had reserved for the
righteous, He rejoiced, for it is said: He rejoiceth at the light of
the righteous. 33 Now Tannaim [differ on the point]: The light
which the Holy One, blessed be He, created on the first day one
could see and look thereby from one end of the world to the other;
this is the view of R. Jacob. But the Sages say: It 34 is identical
with the luminaries; 35 for they were created on the first day, but
they were not hung up [in the firmament] till the fourth day.
36 ..'[171]

[171] http://halakhah.com/pdf/moed/Chagigah.pdf

'All the stars and constellations were created at the beginning of the night of the fourth day, one (luminary) did not precede the other except by the period of two-thirds of an hour.'[172]

Keywords: KaV(line), KaPhuL(bent), CheVeL(cord), KoVeiah(affix), SoL(basket), SeLah(rock), SeDeR(order), YaShaR(straight), ShuRaH(line), TsaV(command),..
Biblical Hebrew Etymology facebook group on כוכב (star):

Jeremy Steinberg: Yerios Shlomo who suggests it is related to כבה (to dim), because the word applies to the dim stars, in contrast to the sun. .. Looks like this is the traditional source of the word כוכב. So says the תולדות יצחק (R' Yitzchak Karo, uncle of R' Yosef Karo) in Ge. 1:16. And I saw a Midrash quoted (couldn't track down the original, it's 2:30am) that the name was given to the (relatively) dim sparks spun off at the time Hashem diminished the moon to its present size..עכב - delay.

RMP: If the moon were as bright as the sun, a molten start? we'd not see stars during night any more than during the day. 'Dim' via CI? as relative size/distance makes light getting here from stars/ planets aside from Sun/Moon pale in comparison to the 2 'big' lights, even if the moon's no longer giving it's own light.

Perhaps Hashem 'held back' the light from the stars by distancing them during the day 4 CI expansion epoch, thus 'dimming' them by distancing/'delaying' their reach here!

Per SPIRAL hyper-dense proto-stars / galaxies were distanced, the 'stretched' heavens during cosmic inflation (CI) day 4, resulting in their light trail channel/lines.[173]

עקב (ankle/angle) also CHaPhuL (bent) touches on how the stellar light-trails are bent due to gravity attraction & magnetic deflection...See SPIRAL on Parallax. ..

[172] www.sefaria.org/Pirkei_DeRabbi_Eliezer.7
[173] T' Succah 22b 'Kochvei Chamah' regarding sunlight.

David Kolinsky: KoKhaBh comes from KBhKBh. The first Bhet drops out in Arabic, Ethiopic, Akkadian and Aramaic, but is preserved in Ugaritic... It is a doubling of the root KVH=KaBhaH which means "to quench / extinguish" but literally means "to press down upon." KoKhaV - star, literally means "an impression".. The transformation from Kaf to Qof only occurs once in the evolution of Hebrew from the word KWH which literally means "to impress a point" and is understood to mean to scald, sear, cauterize." QWH also means "to impress a point" but rather than pressing straight down, the action is with forward motion and therefore the word means "to channel" by which it makes a line. Therefore QW means "line".. In common usage QWH means "to channel" such as MiQWeH = channeling of waters and by extension "to hope or expect something" perhaps a channeling of expectations, such as Tiqweh = hope. With regard to it meaning "to impress a point such as to make a channel" that is a much more difficult proof. /WH = to point to something that exists and want it. aleph becomes Heh >>>HWH = (point to something desired) desiring something (Arabic only). Heh becomes Cheth >> ChWH = to point something out, to instruct. Cheth becomes Kaf >> KWH = (impress a point) cauterize. Kaf becomes Qof or Gimmel >>>> QWH = (impress a point) make a line, channel, expectation GWH = (impress a point) make a concavity, arch, draw inward,..

RMP: SPIRAL may help us understand Hillel and Shammai on what came first, the heavens or the Earth. The 'throne' proto-stellar formation was by/before day two. Earth's super-continent 'the foot stool' established on day three, and 'the set' becomes operational with stellar placement CI day four.[174]

---

[174] T' Chagiga 12a. See T' Brachot 32b for stellar object enumeration.

SPIRAL's 'MVP'[175]
Ours is the Most Valuable and Preferred View in The Universe.

If SPIRAL the observable universe is most likely over 99% of the entire universe. As CMB, arriving here and now at 'c' departure point detectable here and now, is = or greater to the most distant stellar objects.

If SCM-LCDM the observable is under 1% of the entire universe.

If SPIRAL our visible universe has a radius of up to 4B LY max but likely under 2B LY. While distant viewpoints beyond 6K rounded LY see stellar objects within 6,000 LY. So our visible universe is about 333,333 times larger than that of viewpoints beyond 6k rounded LY.

If SCM over large distances there is no preferred view, so the visible universe should approximate the size of ours. So a radius of 46.5B LY.

'Visible (Observable) Universe' defined as the area and mass of the sphere we are by the approx. center of. The radius being the distance to where the most distant stellar object/s are now whose light is at all detectable here and now.

Both SCM & SPIRAL hold that light departed when the object was vastly closer. In SCM up to 46.5-4 = 42.5B LY closer. In SPIRAL up to 2B+/- LY closer.

As it is all for us to connect to Hashem, it makes sense the visible universe is closer to 100%, than it is to 1%, of the entire universe.

For those that want to live in denial of our special place in the universe, holding by the deep-time dependent assumption of 'no favored view over long distances', and up to about 10% is visible, makes sense.

---

[175] For merit for my mother Miriam (Vogel) Pearlman 'MVP' OBM
T' Shabbos 77b R' Yehoshua b. Levi per Rav: 'There's no needless creation/DvaR LeVatalah in the universe'. So visible about whole.

SPIRAL's 'MVP'
The Most Valuable & Preferred View.
Exhibits A – D (not to scale) and guide:

How we know ours is a vastly greater preferred view point in the universe compared to any other distant one.

X= Earth/Sun Ecliptic: Approximate center of the universe.

LY = The distance light travels in one year at light speed.

I = Inner universe. No CE (CR/CB) light from stars therein.

O = The Outer universe starting where I ends.

Y = Radius of "I" less Radius of X in Light Years (LY).

Y = The LY distance from X where I ends and O starts.

Y = Number of years elapsed subsequent to end of CI-CE.

Y predict increases one LY per year.

Y Light Years = age in years of the universe if CI week one.

Y prediction 5,778 + one a year to date as of Sept. 20, 2017.

SPIRAL works if one assumes above 'c' galactic movement away during distancing, or our preference cosmic expansion where the galactic objects stay put in orbital peculiarities and the fabric of space expanded. See our 'magnetic repulsion' hypothesis for the cause and effect mechanism of cosmic expansion. While 2 + magnets now repulse each other into existing space, prior to space, metric expansion of space results.

SPIRAL: the radius of the universe is between 1B-4B LY.

SPIRAL: light limited to 'c' standard light speed 1LY a Yr.

Other factors aside from CE can cause Redshift and Blue-shift. Here we deal with redshift attributed to CE known as Cosmological Redshift (CR) and CB cosmological blue-shift offset due to the hyper dense proto-stars and proto-galaxies expansion during cosmic inflation expansion (CI).

Water which was present prior and during CI, and/or elector-magnetic repulsion, .. may have effected frequency/ wavelength during CI. As did change in density.

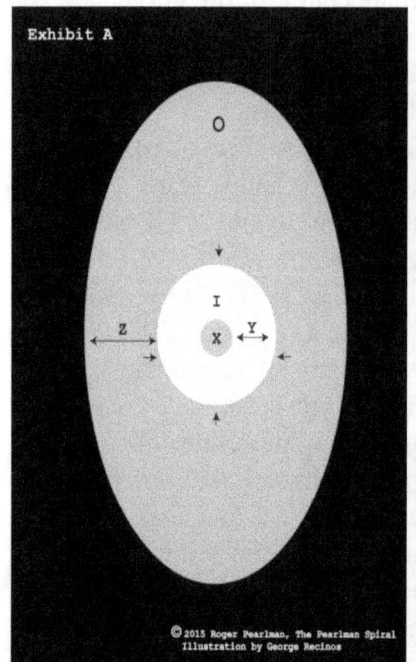

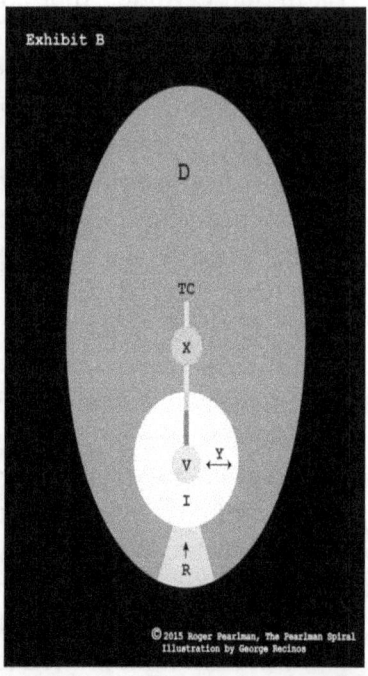

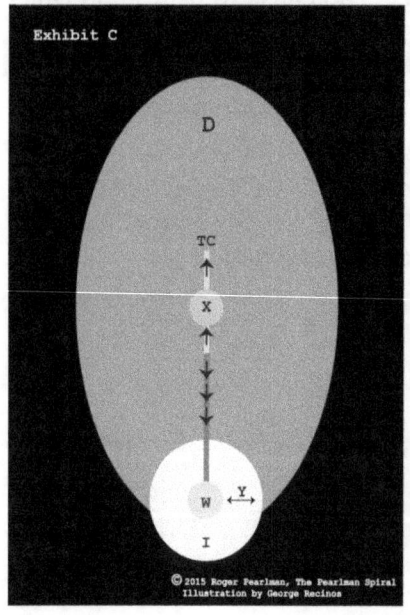

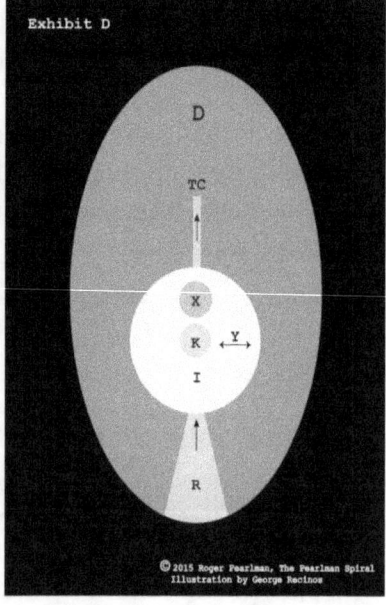

Exhibit A

The perspective of the universe from "X" The Earth Ecliptic.

I = Our Inner universe. Normal light from stars therein. From our observation point on Earth, a band of stars one LY wide should go from CR/CB to regular light over the course of a year, each year. Such band receding from us at one LY per year starting after a hyper expansion inflation CE Epoch.

O = Our outer universe. Starlight from O reaches here   with
      CR/CB. (Cosmological Redshift/Blue-shift offset)

Z = The radius of O the visible universe minus Y. SPIRAL
      predicts this decreases one LY per year.

Y = LY in number = to years elapsed post CI/CE.

CR = is from the stars receding from X on inflation day.

    The Radius of X+I+O is static as a somewhat stable universe. There is no OCE. Light going away from us from the most distant stars will never be visible by us, and few if any stars have broken orbit and are heading out past the outer edge of O, and if so at no greater than 'c' light-speed.

    We see the distant stars in O from the section of their CR/CB light trail formed during CI cosmic inflation day, that if SPIRAL departed the stellar object when the same number of LY away from us then, as the year count since that CI.

    Per the Moshe Emes series this is year 5779 AM, so this year, we are seeing the one light year deep section of all visible stellar objects at and over 5779 LY distant that departed when 5,779 light years distant, so subjected to day four CI expansion.

    Change in density, normal orbital velocities, CBO, gravitational bonding, early water resistance to water, or the like, are variables might mask if/when/where light was exposed to CE.

    If Cosmic Inflation was Cosmic Expansion, the expansion of space, not stellar velocity, caused the cumulative distancing at speeds vastly above 'c' standard light speed. See SPIRAL vs Hubble exhibit CE.

Exhibit B   Perspective from "V"

V = Any fixed viewpoint over 5778 LY to date from X.

I = The inner universe of viewpoint V. Get's regular, not   CE
exposed, light from stars therein.

Y = LY Radius of "I" = to number of years elapsed subsequent to
end of CI-CE.

R = Light with prevalent cosmic redshift from any stars
        beyond Y on the far side of V.

D = Dark, not visible from V, except if w/the same galaxy due to
change in density during CI-CE

TC = not visible from V the column from V almost Y past X,
should have a faint light column which blue shifted, from when V
formed, then stretched to current location, on inflation day,
Applies only if SPIRAL's 'Special Cosmic Expansion'. Ignore if
SPIRAL's regular 'Preferred Cosmic Expansion'.

While we use the term blue-shifted (not CB type) to
indicate light from a star moving in the direction the star is
moving, it could well be non-applicable in a light emitting object
moving away at speeds faster than the speed of light, as is
hypothesized by SPIRAL during cosmic inflation day.

See comments on exhibit C why the light that may have
been blue-shifted heading toward V is overlaid with CR light
starting Y LY past V heading back in the direction of X.

As V originated on the near side of X, about one LY from
X. Each year one more LY of that may have been blue-shifted
column reaches V. Each year the redshifted end extends one more
LY beyond the far side of X from V.

Variables: change in density, water resistance, &/or
elector-magnetic repulsion, increase of oscillation frequency
shortens the light wave-length, so offsets wavelength lengthening
due to day 4 CE, and/or gravitational bonding.

Exhibit C: Perspective from "W" the most distant Galaxies.

W = galaxy at the outer edge of the visible/entire universe.

I = The inner universe of W has regular light only (no CR/CB due to being subjugated to CE) from stars therein.

Y= LY = to years elapsed post CI/CE. Predict 5781 to date.

D = Dark, not visible from W even if W's 'lights are off' for increased visibility. Each year W's light extends one more LY beyond the outer edge of the universe. Define the outer edge of the universe as the most distant stellar objects, not including the 'Y ' LY of their light beyond the edge..

TC = SPIRAL's hyper-dense proto galactic formation preceded hyper CE resulted in TC Light / Radiation trail residue from hyper cosmic expansion CI day 4. Metric expansion of space resulting in W and X to distance from each other. Max cumulative CR due to greatest distance subjected to CE.

'W' sees 'X' when entire universe radius 'Y', when we were 'Y' LY distant, Y years ago.

'Y' LY past X, the part of column TC that already passed us, as 'W' started year 1, 1 LY from X when universe radius 1LY.

Perhaps nearby galaxies (w/in 200M LY) to W, along TC, exhibit nearside cosmic blue-shift offset visible from W.

As all started out in close proximity, w/in 1+/- LY at the start of CI. If they can see back to when they were close to each other as the radiation trails that was elongated during cosmic expansion, with some dissipation over time. So rather than see the entire universe that approximates the visible universe from our central location they see it as a small distant sphere the relative size (depending how far away they came to their resting orbital peculiarity of the sun or a distant star .. with the rest of the universe outside of objects inside their radius 'Y' being dark. See Graph MVP.

See 'SPIRAL vs Hubble' on change in density factor..

Exhibit D:

Perspective from "K", Y-1000 LY from X (the Earth-sun ecliptic), inside our Galaxy calibrate for hyper-dense start..

Y = 5,778 LY to date, + add 1 LY per year.

K = any star/observation reference point now Y-1000 LY from X.

I = The inner universe of K w/ no CR/CB from stars therein.

Y = LY Radius of I – Radius of K = # of years elapsed post CI.

D = Dark, not visible from K even if K's 'lights' are off.

R = CR/CB light subjected to CE during CI from stars beyond Y on the far side of K.

TC = Not visible from K a column that might extend above X represents traces of CR Y LY long. As K formed just below X early on inflation day. By the end of that day it had a redshifted trail from X, Y LY long. Before adjusting for CB.

This Y value will stay constant, as each year "I" increases by one but the top expands by one. The light therein might dissipate as it is absorbed into other matter, or entropy, etc. From this end it is doubtful it is detectable unless you look directly into the column toward K, and K is also visible from X. In general, the further away the star in the outer universe, the thinner the TC trail, all else being equal such as the size of the star, how bright the star is, and the viewing conditions...

Change in density, any water resistance, elector-magnetic repulsion, attraction or gravity... increased the oscillation frequency so shortened the lights wavelength, it offset some of the wavelength lengthening due to day four CI. CI due to hyper intense magnetic repulsion proportional to the hyper-density.

In classic big-bang hypotheses, what caused a big bang & CI, is secondary to the aftermath result. Inflation energy resulted in the CMB. If SPIRAL no OCE, CE coincided with CI inflation day, allowing the expansion of the hyper-dense proto-stars and galaxies. SPIRAL and SCM agree we see CMB now that's come from a LY distance = to yrs. elapsed subsequent to CI.

Exhibits A-D Additional Comments:

For illustrative purposes we are not factoring in stellar orbits, but one should calibrate for them to get the precise actuality in order to predict the next star to lose its CR/CB.

If Y was the depth of the inner universe I was one LY to start Y was less than two LY, as it appeared to Adam during year one AM. Y has been growing a steady one LY per year. So Y and I increased at a high proportional rate to start. About 1/1 yr. one, ½ year two, 1/3 year three and so on.

So the rate of expansion steadily declines, add one to the denominator each year. The numerator being one. So, this year the one LY increase in Y increases Y by 1/5778. The round sphere of I rate of enlargement is also slowing, but at a slower rate.

Other factors can cause Redshift. Here we deal with the redshift known as Cosmological Redshift (CR) attributed to Cosmic Expansion (CE) by the Standard Cosmological Model (SCM) and The Pearlman SPIRAL.

Also not factored into the exhibits is the number of light years (LY) from X that the stars were formed. About one LY is the assumption. As I think Hashem would want the basic constellations to be visible to mankind within a year. Perhaps 40 light weeks worth, like gestation & in the Ark.

It is possible some stars formed during or at the end of the CI, if they formed in "O" then they would not be visible yet from X, unless they formed out of proto-stars that were emitting light prior to departing I. If so we might see a galaxy in O but not it's individual stars if they did not form or start emitting light till they passed the I/O boundary.

So the stars, and or what they are composed of, would have started out hyper-dense then expanded to current size during CI expansion day, to a radius up to about 2B LY deep starting within one LY from X.

Under SCM stars formed relatively early in the life of the universe. As the light from the most distant visible distant stars had to leave soon after the formation of the universe to have already reached us. CMB (cosmic microwave background) readings are evidence of a rapid cosmic inflation CE event.

Hyper-dense Proto-Stellar formation near the start of the universe and a cosmic inflation expansion event. as hypothesized by/in SPIRAL, aligns with Torah and Science.

Snowfall: A natural observation to help conceptualize stellar formation and inflation day. There is great variation with snowflakes, and it is hard to find two exactly alike.

How much more unlikely is it that two stars are exactly alike? Consider that one inch of rain is the equivalent of about 13 inches of fresh snow.

Assume: one cubic foot of accumulated snow having one million individual flakes. Snow falling at an average of two MPH. A cloud height of two miles. With flakes at eight percent water density so that one inch of water is 13 inches of accumulated snow.

The accumulation rate of one inch per hour. So those one million flakes in one square foot would equate to a 12 inch by 12 inch column, 24 miles high of free-falling snow. There are 5280 x 12 inches in a mile.

So the ratio of the height of the stacked static snowflakes, to the height of an equivalent amount of water on the ground is 1,520,640:1.

Now compare the water to the outer band one LY from X that inflated on day four, and the flakes to the stars. Either the distribution of the stars is proportionally sparser, the band's density ratio greater, and or the band was thicker than one LY.

Think of a time lapse film in reverse of a puddle of melted snow, back to when that snow was falling for hours and how high and at what rate it was falling. Then envision when Hashem formed the stars and inflated the universe based on the Spiral Model

I am reminded of the tornado film where a town is assembled from random debris by the forces of 'nature'. While an ID advocate knows he is seeing something in reverse, a NDT adherent who is trying to have a consistent world view, has to consider the storm could really assemble all the structures in the town by chance. :) Even if you slow it down and play repeatedly it will not happen. :)

Other examples to help conceptualize how the highly concentrated outer band filled the heavens: a smoke bomb or gas canister whose fumes fill a room. Bubbles in a glass. Cluster fireworks, Nuclear Energy/Bombs, Sunshine (a little light can displace a lot of darkness) ...

There is a better chance that the stars formed near the central power plant that had to be at/by the initial singularity, pre-cosmic inflation into the void of deep space.

The value of how far from X the outer band/shell of matter from which the stars formed, that per SPIRAL surrounded Earth starting day two, as recorded in Gen. 1:6-7, may be easily determined once we can precisely measure how far it is to where I ends and O starts.

Be sure to consider how our new Proto-Stellar Blue Shift Offset Continuum Hypothesis (CB) that compliments SPIRAL, would impact the amount of, and distance to, visible CR.

On a side note, if a Trillion-fold CI expansion. If SCM 14B years is relative to 5 days per Dr. Schroeder. SPIRAL is more reasonable with A 2B LY radius expansion in one day. With no ongoing CE or missing dark energy required.

Gamma Ray Bursts (GRB) light column issue. Some hold GRB's are responsible for focused light columns. These trails are a lot older, and on average traversed greater distances, under SCM vs SPIRALL. If SPIRAL the light trails formed along the trail from us, the stellar object was moving away from us while being subjected to cosmic inflation CI expansion, until the alleged burst. So no more than 5778 LY and years ago. The closer distance, higher rate of CI means less entropy and less bend, as the higher the speed the more difficult to turn, all else being equal. Perhaps all was not equal as perhaps still relatively hyper-dense.

The higher the speed the less light should be going off the sides of the trail. So overall a lot less energy was required to produce the same factual natural observation. The greater the claim, the greater the burden of proof.

The disproportionate number of GRB's aimed toward us and sufficiently large-scale structures, falsify the Cosmological Principle thus SCM, but align if SPIRAL.

Why are GRB focused light columns to begin with? Is SCM the only option, so why limit consideration to explanations that fit SCM? SPIRAL may hold the best GRB focused light columns explanation.

Superluminal speed, breaking the light barrier further discussion: What if anything special can happen to various objects that accelerate past light speed, or slow back below light speed? Like when a sonic boom occurs when an object breaks the sound barrier. It might depend on the object. Is it light omitting like a star?

Keep in mind light is not subject to inertia, so the type of matter matters:). At certain speeds even non-light emitting objects may emit light. Like comets and shooting stars. Speed, motion, friction, density, are variables. As rubbing sticks vs using flint to start a fire.

If lack of motion equals cold, so motion links to heat/light. So consider any effect if the light trails/columns generated during hyper fast CI expansion, on the hyper-dense Proto-Galaxies and stars.

So a theorized massive star that expired in a GRB might never have been a star at all, but a dense piece of matter that on inflation day left a trail. Perhaps it was consumed by the trip, formed into a star by arrival, or is still in its orbit like a planet or comet.

Density may have been required to keep from breaking apart or even being able to move so fast during cosmic expansion, centripetal force, centrifugal force, and electromagnetic radiation most likely were at play.

Consider Leavening raisin-dough analogy w/SPIRAL. Raisins can represent our hyper-dense proto-galaxies.

As they spaced out they expanded during CI to give the appearance of a black hole in the center of each Galaxy.

If SCM the most distant visible stars now 46.5B LY away, had to form 13B+YA, just after the universe's start.

So even if one uses the deep-time assumptions to believe our universe started 13.8B years ago and the CR is due to OCE, that gives an average expansion rate of the sphere that is our visible universe of $46.5/13.8 = 3.37c$.

If it is true stars either moved or are moving faster than light speed (c), Cosmic Expansion (CE) may be why. If SPIRAL CR is due to past CE, if so no need of OCE.

If the actuality is accelerating CE velocity, over and above a uniform CE rate increase of recession with distance, in all directions, would imply we are a favored center in the universe. See SPIRAL's MVP hypothesis exhibits A-D how we do have a vastly superior optimal view compared to any distant view point., even if CE was at a uniform velocity.

'Distant Object May Explain How Quasars Form and Why None Exist Today' - is best understood in light of SPIRAL hyper-dense proto galactic formation PRIOR to cosmic inflation. Their visible light departed SPIRAL LY radius 'i', 'i' years ago. When we went from hyper-dense to mature size & density.[176]

If the universe is a sphere and not 'flat' and we were not in the center it would be extremely probable we should be moving in the same direction as everything nearby us.

If 'flat' and OCE at SCM's superluminal CE rate the observable universe might be far smaller than what we observe.

Imagine if there was OCE between us and our sun at just over 'c'. The sunlight would never be able to reach us.

The observations align best with us by the center of a universe that is a sphere, where there is no OCE. Every OCE speed and direction would causes new problems under SCM. If it is not the actuality, there is no correct solution within it.

With both natural & man caused motion patterns like dammed up water or a traffic jams, once the blockage is removed the acceleration gradually reaches deeper and deeper into the backup. So if OCE one might predict the inner universe would be shrinking at an accelerated rate. Not increase one LY per year as predicted under SPIRAL.

Talking about patterns. Many man-made designs were inspired designs we observe in nature. Some we can't do as well as the original despite our cumulative knowledge & wealth.

Like the growth through expansion from seed /egg to full grown in plant and animal. Think of a majestic Oak tree that grew from an acorn. The acorn started out at a fraction of it's size, along with thousands of others.

---

[176] Anton Partov's https://youtu.be/q_ZGh33i6Vw presentation on
   https://phys.org/news/2022-09-astronomers-swarm-galaxies-orbiting-hyper-
   luminous.html Nov. / Dec. 2022

Each with enough information and potential to expand a billion times in size, with an underground root system, a canopy of branches several stories high, and the capacity to produce future generations.

Now watch the building a typical two story house. Did the workers not go up & down stairs & hallways. In & out of the rooms many times for many reasons. Framing, wiring, plumbing, insulation, sheet-rock,.. It did not inflate itself, and it does not replicate itself.

Look at time lapse footage not just of the house being built but of the materials being produced from raw material to store shelf. The workers sleeping and eating..

My point is just as the Oak tree grows in an optimal expansion pattern for the result. The light trails visible from the ends of the observable universe can teach us about the optimal expansion/ pattern that resulted in our universe.

If it did not inflate as it did, from where it did, at the speed it did, as recently as it did, and in the manner it did, per SPIRALL, there'd be much more obstruction and poorer visibility.

That we have such optimal viewing, and light from the most distant stars, could not be chance, when considering the context. Just how optimal, see Exhibits A-D then consider the true scale.

Just like with the fossil record, where ID uses the same exact evidence as NDT but derives a higher probability conclusion. SPIRAL has the exact same factual observations as SCM, but proposes a higher probability explanation. If SPIRAL's the actuality, it falsifies all deep-time dependent hypothesis.

If SCM, with its ongoing cosmic expansion and galactic rotation problem, where is all the missing 'Dark' energy and matter? We have looked and the absence is deafening. The natural observations considered via the lens of SPIRALL solves that dilemmas and adds up.

If SPI-RALL even the most distant stars are visible from here, because they formed within one LY from X. A relatively static universe has been in place post day four CI inflation day. If RCCF that was 5,777 years old to date, so predicts Y to be 5777 LY after calibration for CB.

The year lapsed post CI can be independently calculated by measuring Y. So the nearest CR LY distance is an age cap, and adjusted for CB, is = to the actual number of years lapsed.

Per SPIRALL to the outside observer, the rounded distance from X that light extends is the post inflation estimated xB LY + 5,777 LY as light from the far side of the most distant stars moves away from us at one LY per year versus SCM 46B+ LY now after 13B years of cosmic expansion from a post inflation xB LY.

SCM claims the most distant visible stars are now 46.5–4 = 42.5B LY beyond where their light we are seeing now from them departed them VS up to 2B LY if SPIRAL. 42.5/2= 21 rounded = a 2100% greater claim on this one point alone. The time savings is even greater.

The greater the engineer the greater the parsimony in design. Why take 100 yrs. to mfg. a car with a maximum speed of 1MPH that gets 1 MPG if you can take 1 year for a 70 MPH @ 30 MPG? All else being equal.

Why burn down a city to BBQ a chicken when you can do on the grill @ a small fraction of the time & energy?

'Cosmic Inflation Day' day four of creation week, when Hashem expanded / stretched out space. Two days prior to Adam's formation in full stature on day six, which was the first day of the lunar month aka Tishrei aka Rosh Hashanah. Being the start of Year 1 Anno-Mundi (AM). If RCCF the full visible universe is 5778 years old (or young depending on our perspective), as of Sept. 19 2017.

Illustration from 'Blitz' of a fourfold change in density.
Not to scale, as one LY = 63,241 AU.
Baseline parallax 0.5, 1.0, & mature density of 2.0 AU.
Reflecting one stellar object shown at 6k, 12k, & 24k, LY distant,
the increase due to past cosmic expansion.

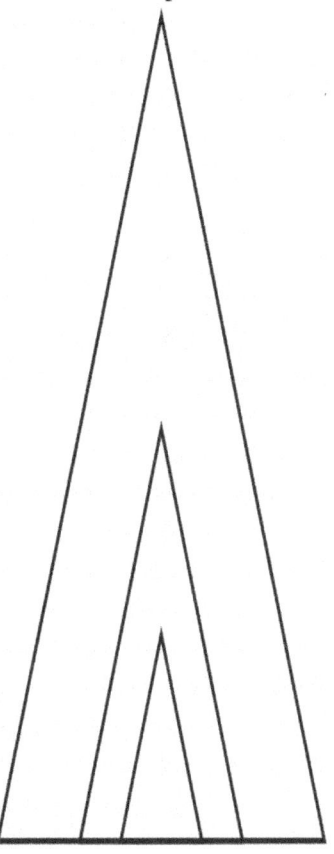

Parallax of stellar objects subjugated to cosmic expansion, reflect
current, not their light departure point, distance.
We observe the light by the baseline. Parallax is constant from a
single stellar object that distanced from the baseline due to cosmic
expansion. We find visible light departed stellar objects over
SPIRAL LY radius 'I', 'I' years ago.

SPIRAL's **"Draw Play"**
Lunar Formation and Original Single Continent Hypotheses
Edition 1.5 dated 21 Iyar, 5777 AM/ May 17, 2017

SPIRAL & SCM, place all matter in close proximity to start.[177] Recent factual evidence indicate the moon and Earth formed from the same 'cloth'. Now it is apparent the moon was 'mixed' here. 'Planetary scientists thought they had explained what made the moon, but ever-better computer models and rock analyses suggest reality was messier than anyone expected.'[178]

Per SPIRAL the proto-luminaries were hyper-dense and preceded cosmic inflation expansion (CI-CE) day four. Lunar cooling from Molten to non-molten surface in even one day is a possibility with expansion and/or water. With expansion think of how much/quickly CMB was cooled by cosmic inflation.

If the moon was molten and x times as dense during the hyper-dense stage of the proto-luminaries by and before the start of CI-CE day 4, if we knew the starting molten temp and know what temp drop is required to go non-larval we can calculate x.

Would expansion cooling even be needed after being drawn up through the day two water? Factor in ocean currents, as we do by the Mabul in The Recent Complex Creation Framework (RCCF).[179]

Think of cotton clothing being hung or steam dried. The flat lunar 'mare' may be a signature of this water exposure. No need of speculated volcanic eruption & filling or billions of years.

The day four CI-CE could be how/when the moons expansion and placement into a relatively stable orbit. Past where the gravitational pull would quickly pull it back to Earth. The escape velocity was far greater than the current minuscule average recession rate. So not an uniformitarian extrapolation.

---

[177] Genesis 2:4 Rashi: all physical matter was created on day one.

[178] Impact Theory Gets Whacked, Daniel Clery, SCIENCE 11 OCT 2013

[179] Link to RCCF and SPIRAL at Researchgate

Perhaps only after the 1656 AM 'Mabul' asteroid impacts year the moon began to recede, perhaps from 400x nearer than the sun? The sun : moon radius of about 400:1 being of interest.

Per RCCF pre-Mabul it was still a single continent thus single ocean. Mabul related impacts and tides may have put us on 'tilt'. So a non-linear Lunar orbit change post-Mabul is reasonable.

If the proto-Sun was used to draw the molten mass that would form the Moon, if from the Earth, it would have been on process day three. Rather than 'spewed' due to rapid spinning as postulated under 'Fission' hypothesis.[180]

To assume more porous than it is can inflate the asserted age.[181] Note how the Lunar lava flows had to be more, not less, fluid due to the weak gravity. Therefore less porous, not more porous. So an age much younger rather than much older, results.[182] The more porous, the more likely it formed here, under stronger gravity, prior to the moon's separating.

Some assume deep-time, extrapolate a uniform meteor impact rate, that may greatly understate the rate, for faulty conclusions, if not deep-time.

Consider how uniform an impact distribution pattern we should predict based on how much time elapsed. Under the laws of probability the more coin flips, the closer to a 50:50 heads to tails ratio we can expect. So all else being equal the larger the disparity between the near side and the far side indicates the fewer the years elapsed. So 'Thousands not Billions'.

We must shield each other a bit, NASA's link claims it's a myth we shield the near side from it's share of random impacts.[183]

[180] http://hyperphysics.phy astr.gsu.edu/hb.../Astro/moonform.html
[181] https://news.brown.edu/articles/2017/03/ina
[182] www.volcano.oregonstate.edu/.../lunar/Overview.html
[183] https://sservi.nasa.gov/?question=3318

Our having about 4x the radius and gravity should attract more meteors. A far greater ratio hit the moon, and with greater force (size & speed), due to lack of lunar atmosphere vs Earth's.[184]

Keep in mind the adage 'when it rains it pours'. It isn't reasonable to assume a constant time interval between each impact, or that each impact is of the same magnitude.

During an asteroid impact here, a thousand times above average debris impacts on a specific region of the moon is reasonable. Think of a meteor shower on steroids.

So it that means the coin is 'weighted' as those impacts would hit mostly one side or the other. Count the number of visible impacts.

Now factor in how many might be associated with anomalies, like asteroid impact/s here (the Yucatan one..) and any known, allow for unknown, there.

So fewer 'flips' and 'thousand not billions[185] of years, can account for the uneven distribution. Account for the softer and dustier near side effect on crater formation. Also when the moon went from being 'alive' to 'dead'. If recent, or if alive now, indicates young.

Another distinction is under SPIRAL's 'Draw' prior to this there was no continent yet as the proto-Earth was covered in water day two. So the proto-moon was attached over the original single continent, and used to draw it, ie gather it up, from below the waters surface.[186]

[184] www.nasa.gov/centers/marshall/news/lunar/program_overview.html#linkI
[185] Term and Title of book coined by ICR rate team Dr. Don DeYoung PhD
[186] Genesis I:9

Then on CI-CE day 4 the sun and moon were placed in their respective orbits. The moon expanding enough to cool to the point it was a reflector, not emitter, of light. If the water and time lapse were not enough to align with the factual observations.

'Inflation energy' may hint how (hyper-dense?) proto-moon, asteroids, stars, planets.. 'stretched out' out on CI day four.

So 'Draw' solves "D" the Earth's orbital plane and "E" Energy in the Earth-Moon system issues that 'Fission' fails.[187]

Thinner crust works just as well with 'Draw' and in no way proves 'filling' by speculated eruptions, speculated years ago, nor does it prove speculated new ones erased old.

'Late Heavy Bombardment', 'Cambrian Explosion' 'punctuated equilibrium', 'asteroid impact', 'meteor showers' attest uniformitarian assumptions are wrong.

Some of this can be tested, the rest the strongest science goes w/ the highest probability. Draw is more parsimonious. Draw appears to be a higher probability.

Either way the higher burden of proof is with a greater claim. Draw makes a lesser claim. So should be the standard default if and until falsified or proven otherwise.

Lunar observations escape velocity, gravitational pull, orbit, tidal forces, recession rate, radio metric dating, fossil evidence... align with ID and YeC not presumption of uniformitarian extrapolation and other deep-time doctrine assumptions.[188]

---

[187] lpi.usra.edu/education/workshops/unknownMoon/How_Did_the_Moon_Form.pdf
[188] http://creation.com/the-moons-recession-and-age Jonathan Henry PhD Danny R. Faulkner PhD article in 'In Six Days' by John F. Ashton PhD http://creation.com/lunar-formation-collision Ronald G. Samec PhD

'The Moon: A Faithful Witness in the Sky' Acts & Facts. 8 (2).[189]

'The 'Age of the Moon' and 'Escaping Moon' by Do-While Jones[190]

Rashi: '..the moon was reduced..' Perhaps as in from emitting light to a reflector of light. When commenting on '..the two great luminaries..' [191]

Now change in density via change in volume, would not just cool the CMB by the end of day 4, and the moon, but could account for many observations, like a rapidly diminished magnetic field on Mars.[192]

---

[189] www.icr.org/article/150/344/ Donald B. DeYoung, PhD

[190] www.scienceagainstevolution.org/v12i9f.htm and v2i2f.htm

[191] Genesis 1:16 Perhaps as in from emitting light to a reflector of light

[192] Minerals also a factor, https://youtu.be/aYpjquXwpH4 Anton Petrov

SPIRAL's 'SNAP'
Proto-Earth origins hypothesis.

The Sun served as a Nest Around our Planet 'SNAP'. Prior to it's being utilized for our anchor, central heat, clock & light.

Think solar-eclipse, the super-hot solar corona light spectrum's evidence of iron. Red-Earth is evidence of iron.

While the corona might be much hotter than the Solar core, there's no reason to preclude the proto-Earth having been by the center of the proto-sun and 'snapped' out via electromagnetic radiation. Like a mom pushing out her newborn, or a modern solar flare.[193]

Even our water formation may relate to the sun![194]

As the water covered Earth was pushed into our orbit with the sun as our center, the gravity of the sun, acted like a magnet at a junk yard, pulling out from the Earth the matter that became the moon. Result: our original single continent, as not everything tugged on broke free of our mass.

So we now have cause and effect of the Earth, moon, original single continent and our early 'red-earth' phase. Also of interest 'Adam' link to 'red' and 'earth' and the 'life is in the edom/red da-am/blood, Adamah/ground'.[195]

Other planets in our solar system may have been likewise formed inside, and pushed out from, the proto-sun. Either shortly before, during or after, the Earth w/moon. The case already independently made by Dr. Thierry De Mees.[196]

Just as escape velocity and gravitational pull attest to ID & YeC and not extrapolation of current trends over deep-time so too do tidal forces & fossil evidence.[197]

---

[193] https://eclipse2017.nasa.gov/origin-corona%E2%80%99s-light

[194] Study Suggests Some Water on Earth Was Produced By The Solar Wind
https://youtu.be/maHTGpUxIWs Anton Petrov

[195] T' Chagiga 12a Heaven (Shamayim) & (water covered) proto-Earth day I? created simultaneously, w/ mixing SHa-Mayim eSH(fire) & Mayim(water).

[196] gsjournal.net/books/Planetary-System-Creation-Theory.pdf Thierry De Mees '04

[197] Danny R. Faulkner PhD article in 'In Six Days' by John F. Ashton PhD

SPIRAL finds the entire universe approximates the Earth-sun ecliptic centric visible universe sphere.

If SNAP is the actuality we literally started within the Sun! The Earth, as defined by our entire orbit, is the approximate center of the universe, the sun, being inside our orbit, our central heating system, and more. :)

Along the lines of SPIRAL's 'Black-hole Illusion Resolution' we view what was, not what is. Perhaps some distant stellar objects were exhausted/spent during Cosmic Inflation Expansion (CI). The visible light trails could be all they amounted to.

'Cochav' star can also be defined as the light trail, not just the object the light trail is/was from.[198]

Think of most shooting stars: Tiny objects, that cause quite a light trail, yet get consumed prior to impact. Now think of light trails up to billions of LY to the end of the visible & entire universe. A parsimonious creation indeed, ideal for the purpose at hand.

With Black Hole Illusion Resolution all galaxies started hyper-dense, so Earth/Sun was by galactic center before expansion to mature size, by optimal place/orbit.

Conclusion: If 'SNAP' by the end of day 3, we were snapped out from our central position in what is now the sun, like a solar flare, by electromagnetic radiation. SPIRAL's 'Draw-Play' & 'Magnetic Repulsion' hypotheses aren't dependent on 'SNAP'.

- RCCF YeC framework: Heat, Time & Pressure: 3 variables in stellar, just as in rock, formation.
- Creation/Destruction 'Processes' may refer to creation of heavy elements during creation week[199]
- 'All was created in full stature.'[200]

---

[198] Talmud Succah 22b 'Kochvei Chammah' regarding sunlight.
[199] Marvin Antleman PhD chemistry on Medrash BR III:7 & Koheles III:11
[200] R' Yehoshua b. Levi T' Rosh HaShanah 11a

SPIRAL compared to SCM-Lambda Cold Dark Matter (LCDM)

Parameter	SPIRAL	SCM
Hyper-dense start, then Cosmic Expansion (CE)	Yes	Yes
Cosmological Redshift (CR) a result of CE	Yes	Yes
CR from net distancing due to CE	Yes	Yes
CR is Evidence of Ongoing Cosmic Expansion	No	Yes
CR is Evidence of Cosmic Inflation (CI)	Yes	No
Universe is 'Flat' on the Surface	No	Yes
CI/CE end: Gravitational Bound Equilibrium	Yes	No
CR attests to Hyper-dense Proto-galactic Formation	Yes	No
Earth approximate Center of the Observable Universe	Yes	Yes
The Observable Approximates the Entire Universe	Yes	No
In Light Years (LY), Radius of Observable Universe	1B	46.5B
Black-holes represent long past hyper-density	Yes	No
CE (CI included) for distancing at over Light Speed	Yes	Yes
Light Speed limited to 'c' Standard Speed of Light.	Yes	Yes
Nearside Cosmological Blue-shift Offset of CE	Yes	No
% Universe look-back to first 96 hours	99.99%	0.01%
CMB is Look-back to very early in universe	Yes	Yes
CMB Near Blackbody due to Cosmic Inflation	Yes	Yes
CMB uniform departure distance. LY traveled:	Radius i	13.5M
'MVP' our Preferred View in/of the Universe	Yes	No
Predicted JWST data: mature galaxies early on	Yes	No
Nearest LY departure distance of any Light arriving now at 'c' ever subjugated to any CE = Radius i	6k rounded	6k to 5M Max
Testable Prediction: Radius I increases 1LY per yr.	Yes	No
Further galaxies: denser, passed radius I earlier, mid-CI.	Yes	No
Every and any vantage point has the same max. Radius i	Yes	No
Requires Missing Dark Energy & Dark Matter	No	Yes
Refutes Copernican + Cosmological Principles	Yes	No
Advantage/disadvantage on: Olbers' Paradox	50k:1	1:50k
// on: Universe volume x density x entropy factor	5Qa.:1	1:5Qa.

Standard Speed of Light 'c' reference points:
Approximate/rounded
Speed of Light =   186k thousand (k) miles per second.
                     11.2 Million miles per minute.
                     671 Million miles per hour.
                     16 Billion miles per day.
                     6 Trillion miles per year.
Earth: Radius 4k miles. Circumference: 25k miles.
Earth to moon: 235k miles = 1.25 light seconds
Earth to Sun: 93M miles = 8.3 light minutes = 1AU
The distance between us, the moon and the sun varies. Sun 400
times (rounded) further & larger than the moon.
A light year is about 63k times the distance to the sun, so about
25M times the distance from here to the moon.
Astronomical Unit = 1AU = 93M miles = Sun to Earth
A Parsec = 3.26 Light Years (LY) = 648k/pi AU
So light travels 0.307 Parsec's per year.
Nearest star, not the sun, Proxima Centauri @1.3 Parsec.

Pearlman SPIRAL vs Hubble
On the increase in Cosmological Redshift with Distance

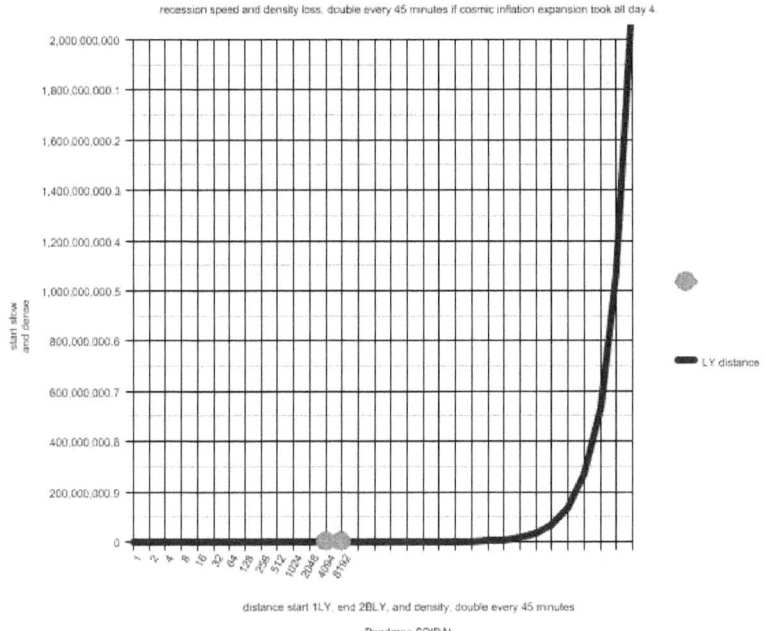

distance start 1LY, end 2BLY, and density; double every 45 minutes

Pearlman SPIRAL

Exhibit CE

Note how hyper-dense proto-galactic formation is a far greater factor when Cosmic Expansion (CE) factored in. As the recession velocity (that correlates to the change from hyper-dense to mature density) does not pick up steam till well past ¾ through the cosmic inflation expansion.

So all distant stellar objects we see here and now. from when the nearest distance to those ever subjugated to CE, assume by the circles by 6k LY, were receding much slower, and were much denser than average.

Assuming a uniform expansion rate doubling 32 times over the course of 24 hours. Starting at a radius of one LY and ending at 2B LY. So doubled every 45 minutes. As the universe reached a 6k LY radius velocity was reaching 6k LY per 45 min. So by the finish almost 2BLY per 45 minutes.

Now 2B divided by 32 = 62.5M. 6K /32= 187.5, so 62.5M / 187.5 = 333,333.33 faster on average over the course of CI-CE, than the speed it was receding, at the light departure point, of 6k rounded we see all the more distant stellar objects from, here and now.

Keep in mind the classic raisin dough analogy. It takes the same time to double from one LY to 2 LY radius as it would to double from 1B LY to 2B LY.

The universe as a whole, individual galactic and individual stellar, density, all having an inverse correlation to the metric expansion of space.

So over 99% more dense on average when the universe had a radius of 6k LY, than at mature density. Mature density reached at the end of the day four Cosmic Inflation CI-CE epoch. When the universe at a radius estimated at up to 2B LY.

Thus a lot brighter, and tighter, light trails from distant stellar objects, than one might expect under SCM-LCDM assumptions.

If SPIRAL no CE subsequent to CI, so cap at 4B LY radius max, but consider the possibility of a much smaller radius. Test for values as low as 60M? LY.

While we assume a steady doubling of size and reduction in density, over 24 hours, test for other spans like 12 hours to see if they fit the measurements of the empirical observation better.

While we assume brightness correlates to density, all else being =, other factors may come into play. For example at what point, and to what extent, might gravity limit the amount of light emitted. Think of the hypothesis a 'black hole' is the result of light unable to escape hyper-density.., then consider our 'black-hole illusion resolution'.

Draft 2.1 Dated: August 6, 2018 / 25 Av, 5778:

Do Cosmological Redshift (CR) observations align as well or better with the SPIRAL cosmological redshift hypothesis and model or with the current consensus LCDM-SCM model?

Both models agree there is prevalent cosmological redshift of visible distant starlight, that overall increases with distance, starting relatively nearby.

By about 1 Million (M) Light Years (LY) stellar objects light begin to manifest cosmological redshift. By definition means that light was subjected to cosmic expansion (CE).

If SPIRAL that exposure to CE was in the past, during the cosmic inflation (CI) cosmological expansion epoch, which was relatively early in the history of the universe, and no subsequent CE. If SCM that exposure to CE was subsequent to CI, and the CE is ongoing.

We hope to provide the equation that would best reflect the measurement of CR if SPIRAL and compare it to what one should predict if SCM. Then consider if the Hubble equation best reflects what one would predict if SCM, or if it twists the observations to fit the hypothesis of ongoing CE.

CR level factors and variables.

- Distance 'D' over which the light was exposed to CE should also effect the amount of CR as we hope to explain. 'D' = 'V' time 'T'.
- Velocity 'V' the speed of the cosmic expansion, as a variable in CR is well established. Distance of the CE / time the CE lasted = V.
- Time 'T' the duration of the cosmic expansion.
- Change in Density 'C' on departure is a material consideration as we will explain..
- Luminosity of stellar objects: Use a fixed output per fixed time, regardless of CE velocity, or any change in density. Later one can test w/ higher and lower values.[201]
- Cosmological Blue-shift (CB) Offset (CBO) in SPIRAL we explain why one would predict CE not only resulted in CR but also some CBO).
- Depending on the starting density of each Galaxy, it's ending size, shape, orientation and distance.[202]
- Normal orbital velocities can cause red-shift and blue-shift. Overall this factor diminishes with distance as CE increases with distance.
- SPIRAL hyper-dense proto-galactic preceded CI-CE is the SPI in SPI-RALL.[203]
- To learn all the science, one must study and fairly consider the density variable.
- This is even before taking into account exhibit CE

---

[201] https://en.wikipedia.org/wiki/Luminosity
[202] See SPIRAL's 'Cosmological Blue-Shift Offset' hypothesis.
[203] SPIRAL cosmological model amazon.com/dp/1517214122

Cosmological Redshift (CR), SPIRAL vs SCM.

If SPIRAL The universe started by the near singularity, and was still hyper-dense. Estimate for now a radius of 1 LY +/-, by the start of the CI-CE epoch.

SPIRAL explains how/why the Light Year (LY) distance at departure to nearest visible light here now, that was ever subjected to any Cosmic Expansion (CE), caps the LY radius of the entire universe, when that light departed.

To know if light was subjected to CE consider change in density 'C', CR, and Cosmological Blue-shift (CB) offset (CBO) as if SPIRAL, it was not just the universe subjected to CE, but every galaxy too.

The LY distance to the nearest departure point of light subjected to CE we will call ' I ', for the radius of the 'inner' universe from which we only experience regular light, not light subjected to CE.

Assume one year post CI the LY radius of ' I' was 1. After 100 years 100 LY, after 5778 years it would be 5778 LY to 'I'. For simplicity we will omit the 1+/- LY start point, as relatively immaterial.

While about 1M LY rounded is the cap to ' I ', CBO, change in density, or early resistance such as if water, would mask the effects of CE.

So 5778 is already within that cap, once the science advances, by factoring in 'C' and CBO, the estimates for ' I ' distance may get a lot closer to our perspective of 5778 LY to date (and increasing 1 LY per year) being the actuality.

If SPIRAL the LY Radius of the sphere that is the visible universe is about 2B LY. Till we can pin down more variables we suspect a 70% probability the radius is between .05B LY to 2B LY, a 20% chance under 500M LY and a 10% chance from 2B LY to 4B LY.

The 4B cap as, per SCM, no visible light here now, departed from beyond 4B LY. The visible universe is only assumed to be larger due to the assumption of ongoing cosmic expansion. As if the most distant starlight departed within 4 B LY as maintained by SCM.

Stellar objects being up to 13B rounded yrs. old as per SCM, & if there were no ongoing comic expansion, there would have been no more cosmological redshift for the past 9B years and it would be obvious the universe has a stable radius up to about 4 B LY.

So if the CR is far better explained by no ongoing cosmic expansion, there's no reason to suspect the visible universe is larger than 4B LY max. So for now assume 2B LY radius of the visible universe, but do not preclude values from 0.5B LY to 4B LY.

If SPIRAL assume the CI-CE took 24 hours. Later consider longer and shorter time spans. So the uniform CE resulted in the recession of the most distant visible stellar object 2B LY in 1/365 of a year.

After which there was no ongoing cosmic expansion. So cumulative CE over the 2B LY distance in 1 day adds up to 730B times standard light speed (c). Link here how that compares to expansion rates if SCM.[204]

[204]www.esa.int/Our_Activities/Space_Science/Planck/The_cosmic_microwave_backg round_and_inflation

Later we will see why a decrease in density with distance, greatly offsets the increase in CR with distance had there not been a decrease in density.

The variables that may have played a role, prior to the arrival of light may be indistinguishable to the observer here. See our rubber-band & TP examples.

Context provided by each models presumptions, help determine the variables. Such as change in density, CB offset, CE distance, velocity and time.

For example in rock formation and stellar formation, heat, time and pressure are 3 variables. When two variables are known, it helps determine the third.[205]

If SCM the rate of cosmic expansion CE averages 3.27 (c) for the past 13B years, so that the light visible here/now from the most distant visible stars, departed them when they were up to a maximum 4B LY max distant and are now up to 46.5 B LY distant.

So they have been subjected to up to 46.5B-4B=42.5 B LY of CE. The light here now having been subjected to 13B LY traveled less 4B max start point = 9B LY of CE

For our calculations and comparisons assume the CE was at a uniform rate over the entire sphere that is the visible universe, we are by the center of. Starting at the 'I' the distance to the nearest known departure point distance of light reaching here now that was ever subjected to CE. Later you can try alt variables in case the CE wasn't uniform.

---

[205] 'RCCF' framework of six principles for understanding science. www.amazon.com/dp/B077Q4KB9V

The Copernican Principal, premise assumption premise of Modern Cosmology & SCM. If valid the view from any distant viewpoint should see the same increase in recession with distance as here. That means the CE has to be fairly uniform over long distances throughout the universe.

If the actual CE is faster in distant locations from us, so not due to more cumulative CE with distance at a uniform rate, would suggest we are by a/the preferred center.

'..ON the EXPANSION of the UNIVERSE ..

It was Hubble himself who correctly explained the red-shift as indicating that distant galaxies were radially moving away from the Earth. In every direction, these vast accumulations of stars and interstellar matter were moving outward at enormous speeds.

He called this motion, recession. He showed that the velocity of recession was greater at greater distances. He also showed that this particular type of recession was consistent with the concept of a general expansion of the universe. We will examine this last point in more detail in just a little bit. ..

After observing many galaxies, Hubble was able to quantify his results. His law of cosmic expansion states that an observer at any point in the universe will observe distant galaxies receding from him/her with radial velocities proportional to their distance from the observer.' .. '[206]

---

[206] grc.nasa.gov/www/k12/Numbers/Math/documents/ON_the_EXPANSION_of_the_UNIVERSE.pdf

If SPIRAL we have already falsified the Copernican Principal by showing how/why we by far have the optimal view, of a far higher percentage of the visible universe, than any other distant viewpoint.[207]

If SPIRAL every point in the universe gets 'normal' light, ie light that was never subjected to CE, from light that departed w/in the LY distance = to the number of years elapsed subsequent to CI_CE which with RCCF framework for understanding we can predict will one day be established at 5778 to date + 1 LY per yr.

With Cosmic expansion (CE), it is not the stellar object moving, but the space between us expanded that is distancing us. So, the cumulative effect of the CE over the distance the object recedes due to CE can add up well above the speed of light.

Now assume regular light departs a relatively stationary galaxy when it was 1M LY distant from our viewpoint. That it is subjected to 3M LY worth of CE along the way so will not arrive till 4M years from now, as it has had to travel 1+3=4M LY to get here.

Now another galaxy 50% closer to start we see in .5+1.5 = 2M years. All else being = , the normal light as it departed the similar object, both get subjected to the same uniform rate of CE on the way here, the only difference being the distance, subjected to that uniform rate. We know the greater the distance the greater the CR. So we know 'D' distance effects CR,

[207] See SPIRAL's 'MVP' hypothesis

'V' velocity should not matter as long as enough 'T' time and visa-verse. As D = V times T. Distance of CE subjugation is primary. If we don't know the distance and we know V and T we can derive D. With D we do not need V and T to predict CR.[208]

One way to think of it. Take a rubber-band stretch it in two in a second. Now stretch it in two in ten seconds. It should end up appearing about the same either way. So too the amount of CR should measure about the same, after being subjected to the same gross amount of CE had sudden or gradual.

After 1B LY subjected to CE rate of 4(c) or 2B LY at 2(c) or .5B LY at 4.5(c). In each case we look at the difference between the CE rate and 1(c).

If the rubber-band can double a few times great. If not try the TP prop. Keep in mind the law of conservation of energy. The rubber-band has no more and no less matter if stretched 2x or 10x. So too there is just as much total light from a stellar object emitted in 1 day if elongated over 2LY or 200LY.

Using the sun to the Earth to illustrate. Use 8 light minutes rounded as the time it takes for light at the standard speed of light 'c' to travel the distance from the sun to the Earth, aka 1 Astronomical Unit (AU). [209]

Think of a light emanating from the sun to Earth starting now and for the next 8 minutes. In 16 minutes that light, and the light already in transit will have reached us. So, 16 minutes of sun light, luminosity.

[208] See SPIRAL's CREST hypothesis.

[209] http://units.wikia.com/wiki/Parsec link explains measurement units definitions...

Now if the distance to the sun contracted via cosmological contraction at 'c' we would be at the sun in 8 minutes and we would get 16 minutes of sun, as we would receive all the light already in transit plus the 8 departing during the 8 minutes.

If the contraction was at 8(c) we would get just the 8 minutes of light in transit + 1 minute of light = 9 minutes of light with 7 minutes to spare.

Likewise, if the distance between us and the sun, increases due to CE by/before the light is in transit it will take longer than 8 minutes for the fixed amount of light to arrive.

If it is subjected to CE it is 'stretched' out and takes longer to reach the fixed viewpoint. If subjected to cosmic contraction it will take shorter.

Another way to try and visualize, take a full closed roll of toilet paper and estimate how long it takes to move across your table or room at 1 inch per second in place of 'c'.

Now accordion fold, to 2, then 4, 8, 18, 32,,. times the length of the roll, and calculate how much longer it takes for the entire roll, to depart one end of the table and completely go off the observer end, at the speed limit.

Imagine the full bunched roll accordion fold is normal light, and expanded lengths the same light whose frequency has been elongated, so CR'ed as subjected to CE. The longer the length the greater the CR.

Try at various rates of CE and distances subjected to CE. Now roll it back as fast as you can, as we don't want to waste paper and may need it later.

Now with that last experiment, like the classic raisin dough one, think how the further from the observation point the greater the CR. As each time the distance from the observer doubles, due to the cumulative uniform CE, so the increase in CR per fixed distance of CE slows.

So each % increase in CR means a like % increase in D. From the nearest distance to the start of CR, to the most distant visible object/s.

The estimate for said distance depends on the model. The light in transit was subjected to a uniform rate of CE. Have 100% CR corresponds to where the model represented would predict the most distant visible stellar object's are now minus where the CE began.

So if 0 CR is under 1M LY, 10% CR is about 46.5-4=42.5(0.10) = 4.2B LY if SCM, 200M LY if SPIRAL. 100% would be about 2B LY if SPIRAL and 46.5-4 =42.5B LY if SCM.

If SCM try the alternate 13-4 = 9B, the max LY CE distance the most distant visible light was already subjected to.

If CR is a function of CE distance exposure, that is a awful lot of D if SCM even using 9B LY instead of 42.5B LY. SPIRAL with 2B LY and change in density offset factor may be the far better solution.[210]

[210] www.astro.cornell.edu/academics/courses/astro201/hubbles_law.htm

Pearlman SPIRAL equation for Cosmological Redshift

$$CR = D / C / L$$

CR = Cosmological Redshift = Redshift due to CE. On a scale of 0-100% with the least CR by the nearest stellar object, whose CR light arrives here/now. 100% CR being from the most distant visible object. In 20 increments each adding 5%.

CE = Cosmic Expansion, Expansion of Space.

'C' = Change in Density: If SPIRAL the distance to the galactic center at the end of the CI-CE divided by 'I' the Light Year (LY) distance to the nearest galaxy whose visible light here/now was subjected to CE on departure. 'I' is = to the LY distance = to the number of years elapsed subsequent to the CE-CI event. 5778 years & LY, based on the tightest chronology and SPIRAL predictions. If SCM no change in density, use a factor of 1.

'D' = Distance: Cumulative distance of CE transpired between us & a stellar object, in LY. Start by the nearest stellar object whose CR light arrives here/now. End by the most distant visible object. If C and L values are one, estimate D doubles with each of the 20 5% increases in CR. D = V (T)

'L' = Luminosity generated by the object in one year.

For SPIRAL estimate time of CE exposure = 1 day, so multiply by 1/365. If SCM use factor of 1.

'T' = Time of CE subjugation in years. T = D / V

'V' = Velocity of the CE. V = D / T

Use LY distance to a galactic center, which should be the approximate average of the CR + CB of individual stellar objects therein, if not factoring in Cosmological Blue-shift (CB) offset of near side, & CR enhancement of far side, stellar objects therein.

Using the average CR of each galaxy by their center, also often helps reduce static from normal orbital velocities of the stellar objects within said galaxy.

SPIRAL's: 2B LY 'D' of CE, 346,141 'C' in 1/365 years:
T = 1/365 of a LY it took that galaxy to recede that distance during the cosmic inflation (CI) CE epoch.
V = 730B(c) = 2B(365) of CE then divide by C = 2B/5,778 = 346,140.53 = 2.11M so about 1k fold less CR build at 100% CR, then if plain 2B LY 'D' before luminosity and change in density factors.

100% CR defined as the highest galactic average level of CR, SCM and SPIRAL predict to be by the most distant visible galaxies.

SPI of SPI-RALL = Hyper-dense proto-galactic formation preceded that CI-CE and no subsequent CE.

CB = Cosmological Blue-Shift offset. While galactic centers were by the end of CI-CE at distance determined by the velocity and time, stellar objects in each galaxy need to be calibrated for the expansion w/in the expansion. As each galaxy expanded on a micro level from hyper-dense to mature during the macro CE of the universe as a whole.[211]

---

[211] SPIRAL's 'Magnetic Repulsion' and 'Black Hole Illusion Resolution' Hypotheses.

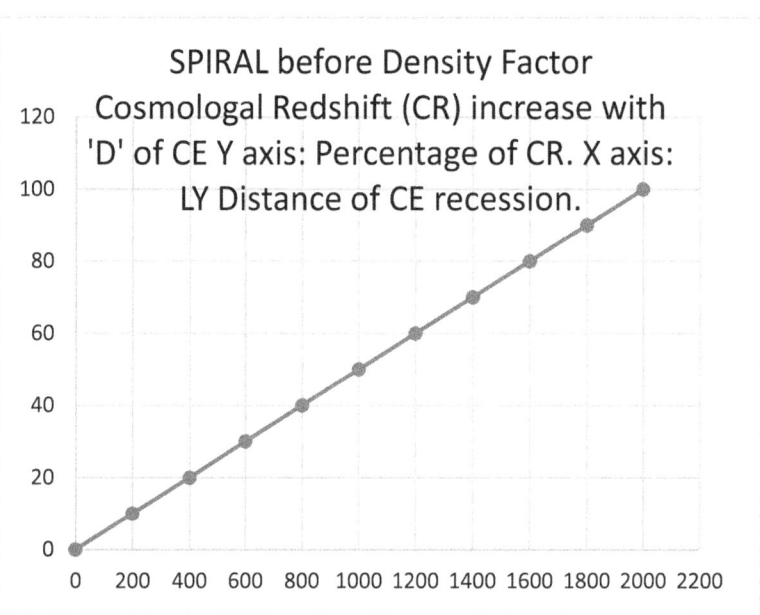

SPIRAL before Density Factor Cosmologal Redshift (CR) increase with 'D' of CE Y axis: Percentage of CR. X axis: LY Distance of CE recession.

$X = Y \% (2B\ LY)$

SPIRAL before density factor.

Increase in Cosmological Redshift (CR) with increase in 'D' distance of Cosmic Expansion (CE).

$Y = \%$ Cosmological Redshift (CR) from 0 to 100

$X = LY$ distance to galactic centers, starting from 1M LY the rounded up distance to the nearest object whose light may have been subjected to CE, to 2B LY.

$X =$ in millions (M) of Light Years (LY) so $0.10 = 100k$ Thousand (k) LY, $1.0 = 1M$ LY, $1k = 1$ Billion (B) LY, $2k = 2B$ LY.

To galactic centers, to average out the near side CB offset, and enhanced far side CR, of individual stellar objects.

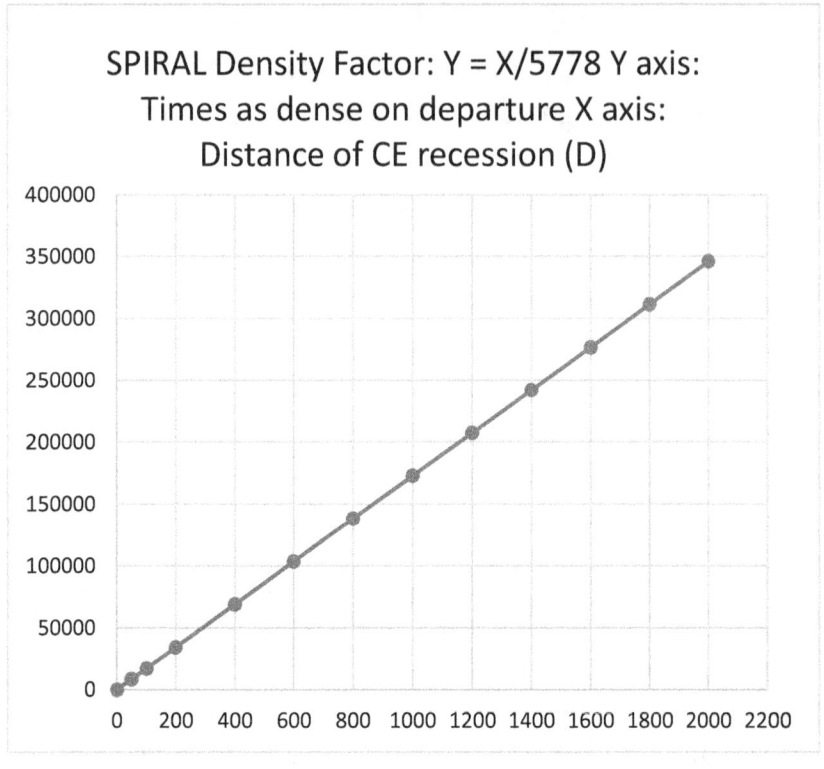

SPIRAL Density Factor: Y = X/5778 Y axis: Times as dense on departure X axis: Distance of CE recession (D)

Chart Title = Increase in 'D' distance in LY of CE w/ decrease in density 'C' if SPIRAL

Y = Times more dense on visible light here & now departure than at stellar objects mature density at end of CI-CE epoch

X = LY distance stellar object was subject to Cosmic Expansion up to about 2B LY if SPIRAL

X starts at '0', which is under 1M light years from us the rounded up distance to the nearest object whose light was subjected to CE.

X = In millions (M) of Light Years (LY). So 0.10 = 100k LY 1.0 = 1M LY. 1k = 1 Billion (B) LY. 2K (thousand)= 2B LY..

Luminosity factor, not reflected in this density chart exhibit.

If SPIRAL: Hyper-dense proto-galactic formation decrease in density, over the course of CE slows, by helping offset, the increase of CR with 'D'.

CR Light here/now departed when 5778 LY and 'D' at the end of CE 1M LY 173, if 1B LY 173k, times as dense on departure.. Divide 'D' by 5778 to get 'C'..

Hyper-velocity of the Cosmic Inflation CE 1/365 luminosity output appears as an increase in CR build. So divide by 365 to get 'C' net of the Luminosity factor:

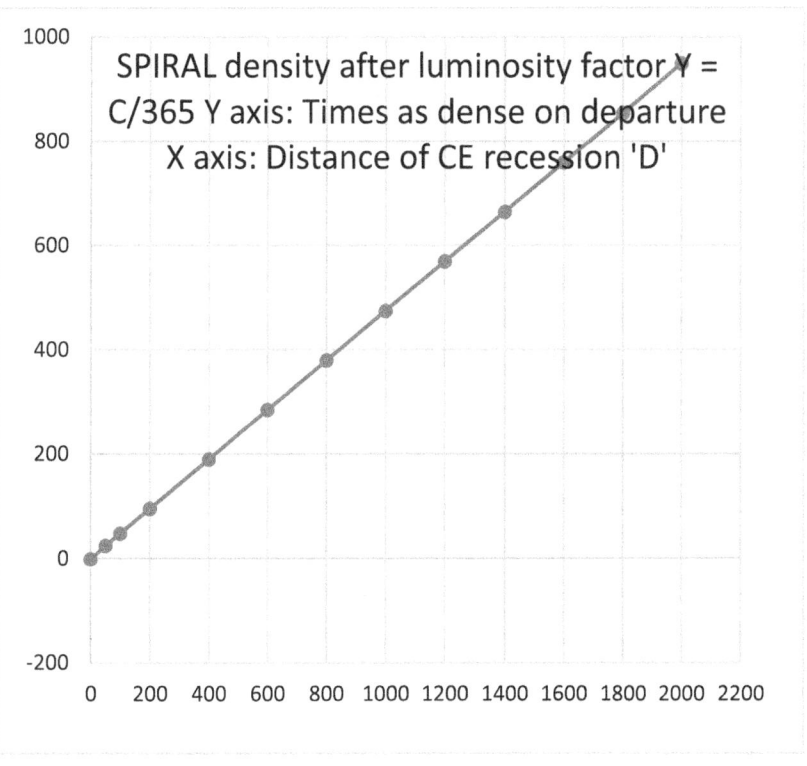

X axis	Y axis
LY 'D'	greater density
in millions	on departure
0	-2.10
50	23.77
100	47.42
200	94.83

400	189.67
600	284.50
800	379.33
1000	474.16
1200	569.00
1400	663.83
1600	758.66
1800	853.50
2000	948.33

Like the graph predicting what CR vs 'D' values should measure if SCM, prior to normal orbital velocity consideration.

Try a start point of one LY from our observation point. The end point is where a galactic center ended up at the end of the CI-CE event.

So D = distance is velocity 'V' up to 730B times T (time) = 1/365 = up to 2 B LY.

If SPIRAL a galactic center now 2B LY distant, the approx distance attained at the end of the cosmic inflation expansion event (CI-CE), we are seeing from when the light departed it 5778 LY distance, 5778 years ago when it was 2B/5778 = 346,140 times as dense as the galaxy was by it's mature density at the end of the CI-CE event.

So a galaxy 1M LY distant we see here now the light that departed it when it was 5778 LY distant and 173 times as dense. As it was when it reached 1M LY at the end of CI-CE.

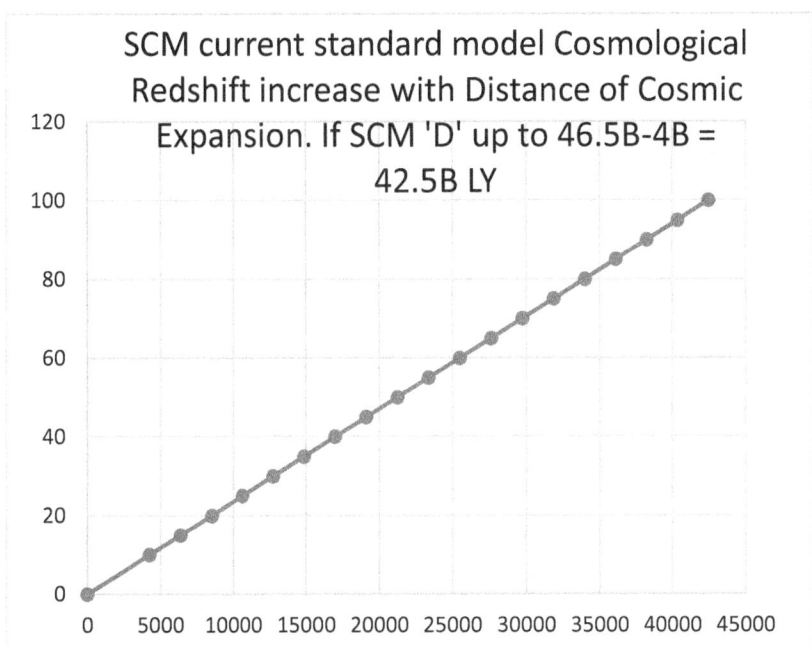

SCM current standard model Cosmological Redshift increase with Distance of Cosmic Expansion. If SCM 'D' up to 46.5B-4B = 42.5B LY

SCM if real.

X = Y%(42.5 B LY)

Chart Title = Increase in 'D' w/ increase in CR if SCM
Y = % Cosmological Redshift (CR). Start 0, end 100.
X = LY 'D' Distance of Cosmic Expansion (CE) 46.5 B – 4B=
42.5 max to date, (try with 13-4=9B too) if SCM
X starts at 0 which is 1M rounded light years from us.
X = in millions of LY so 1.0 = 1M LY, 1,000 = 1 B LY ..

If SCM up to: 42.5 B LY D of CE / T = 13B years so
V (Average Velocity) = 3.27c, or 13-4= 9B/13B = 3.25c
C as in change in density is non-applicable if SCM.
CR of up to 3.27(13B) = 42.5B LY worth of CE.

Everyday example of velocity times time = distance. If travel at: 60MPH for 1 hr. the distance is 60 miles. If 24 hrs. 60(24) = 1,440 miles.

So if velocity is 60M LY per hour then after 1hr. = 60M LY. If 24 hours 24(60M) = 1.44B LY.

Assume a uniform cosmic CE expansion rate over the distance during CI-CE.

I LY = Distance light at c travels in a year. So 365 'Light Days' distance,

1 Light Day = 180 times the distance from here to the sun 60(24)/8 (rounded) = 180 AU

1 AU (astronomical unit) = Distance from the sun to the Earth, to go at light speed 'c' takes 8 minutes rounded.

C = Light speed, small c is the standard speed of light.

C = Change in Density. Per SPIRAL, Proto-galactic formation was hyper-dense and by the start of CI-CE. By the end of CI-CE the Galaxies were their mature 'normal' densities. The greater the end distance the faster it receded (based on a uniform CE rate regardless of distance). So further away when it reached the same density of a similar galaxy that was closer to us..[212]

Graphs predict what CR vs 'D' values should measure prior to normal orbital velocity consideration.

---

[212] SPIRAL's 'Jiffy Pop' with 'Electro-Magnetic Repulsion' hypothesis. On a galactic level each kernel a stellar object. On a macro level each kernel a galaxy.

How SPIRAL and LCDM compare on the Cosmological Redshift (CR):

SPIRAL up to 2B(365) = 730B/346,140 = 2.1M rounded max net equivalent subjugation at galactic centers. Less on near, and more on far, galactic sides due to CBO.

Thus we might expect with the more distant visible stellar objects to see the collective Galaxy as a whole rather than the million/s of individual stellar objects that comprise it.

As light reaching us at any instant from 2.1 million stars subjected to 2.1 net B LY worth of CE frequency elongation luminosity equates to about the normal light of one local star whose luminosity is the average of the 2.1 distant ones had their light never been subjected to CE.

CR = Distance of CE / C change in Density, before calibration for Cosmological Blue-shift offset and normal orbital velocities. Use annual luminosity output.

SCM values if CR = based purely on amount of CE the most distant visible stellar objects were subjected to, as no change in density if SCM.

If we use 46.5-4=42.5B LY 'D' for SCM:
42.5B / 2.1M (rounded net for SPIRAL after density and luminosity factors) = 20,238.10

Even if instead of SCM's projected 46.5B LY less up to 4B LY start point = 42.5B LY CE subjugation we use the 13B LY travel distance = 4B departure + 9B of CE subjugation of the visible light here /now we derive:
13B LY – 4 BLY = 9 B LY vs 2.1M if SPIRAL.

While CR if SCM is up to 4,268 times greater at 100% CR if SCM than if SPIRAL. 9B/2.1M = 4,268 greater frequency elongation at 100% CR due to CE if SCM than if SPIRAL. This means SCM may not be viable.

Another way to derive the 4,268 approximate times greater net subjugation to CE if SCM than SPIRAL based on up to 9B LY 'D' of CE if SCM and 2 B LY'D' if SPIRAL is 9/2 = 4.5times the distance 'D' if SCM.

After factoring in luminosity and density factors that apply if SPIRAL the light visible here now departed when the most distant visible object, was 948.33 as dense. 948.33 (4.5) = 4,267.49 times greater CE subjugation if SCM vs if SPIRAL.

Again using 42.5 B LY CE 'D' if SCM based on were the most distant visible objects are now predicted to be if SCM: up to 42.5B / 2B (if SPIRAL) = 21.25.

948.33 (21.25) = 20,152.01 times greater CE subjugation claim if SCM vs SPIRAL requirements due to Luminosity and Change in Density factors.

Measurements of the increase of CR in distant stellar objects with distance do not fall exactly along a narrow line.

Like SCM with SPIRAL we recognize not all red-shift or lack thereof, is due to Cosmic Expansion.

Normal orbital velocities during the light departure toward us shorten, away from us elongate, the lights frequency. Red-shifted light's frequency being more elongated then normal light. Blue-shifted light a shorter frequency than normal light.

Overall the greater the galactic distance, the greater the CR, the less material effect regular orbital velocities should have all else being =.

SPIRAL benefits by consideration of and accounting for Change in Density, and Cosmological Blue-shift offset (CBO) of near side stellar objects. Hand in hand with CBO is enhanced Cosmological Redshift (CR) of far-side, stellar objects.

When it comes to understanding and explaining stellar objects that are distant enough for their visible light to have been subjected to CE.

For example if CBO a group of stars by the center or on the far side of a galaxy 10.1M LY away could easily be more CR than stellar objects on the near side arms of a similar galaxy 10.2M LY.

So this should widen the line to a band, what we would predict for the level of CR of stellar objects. With the average CBO near side and far side in each particular galaxy, averaging out by each galactic center. Just like snowflakes, clouds, rocks, trees.. outside the lab, expect a spectrum of variation. So with stellar objects, expect few, if any, to be identical.

The amount of near side Blue-shift offset or enhanced CR on the far side, depends on the size, shape, orientation of a Galaxy and how far the stellar objects are from their galactic center. and the distance.

Like normal orbital velocities, how material an effect CBO has on the CR measurements diminishes with distance.

The change in density is a major factor that like the CR increases with distance, so has a material role in reducing the CR all the way till the light goes off the visible spectrum.

Without the hyper-dense start and change in density offset, a good part more of our visible universe might be past the visible spectrum.

The way the universe galaxies started and expanded has the added benefit of a much higher % of the entire universe is visible if SPIRAL and a much crisper night sky if SPIRAL.

Think of Olbers' Paradox, and how hyper density, greater velocity, lead to crisper light trails with far less scattering of the light if not from a hyper-dense start and or over deep-time.

By definition all else being =, CR light with it's elongated frequency, is not going to spread/scatter so light up a greater field of vision, as = amount of regular light.

Think of the example if the sun receded from us at .5 light speed due to cosmic expansion (CE). It would take more than 8 minutes for the light fixed amount of light in transit to reach us.

Now consider starting a new universe, completely dark with just us and a sun about to turn on. If it was moving toward us or away, if receding or drawing near by cosmic expansion of contraction. A rubber band, pleated curtain, napkin, or a roll of TP folded like an accordion can help one visualize. ...

Notice how they stay the same mass if stretched or bunched. So too a fixed luminosity output, is constant. Think of regular mid-range light frequency.

If the frequency is/was elongated over x LY of CE, the greater x is/was the greater the CR.

So too, if the frequency is/was shortened via cosmic contraction, the greater the contraction, ie shortening of the distance, the greater the cosmological blue-shift.

When it comes to how much CR the visible light arriving here & now should exhibit, what we need to know is the CE distance. Unless we know both the time lapse and the velocity of the CE the other is inadequate.

As when we look at the rubber-band, TP, .. or CR'ed light, on arrival, it will look the exact same if it was subjected to 1B LY of CE in one year at the velocity of 1B (c), as it would if the velocity was 1(C) over a billion years..

Does the Hubble equation uses velocity to bend the predictions under ongoing CE hypothesis? It masks the massive distance of CE exposure if SCM..[213]

As we saw above SCM makes a claim on the true amount of CR one should predict if SCM (which should be based on the CE recession distance 'D' as if SCM no 'C' change in density and no Luminosity adjustment) .

The greater the claim the greater the burden of proof. If SCM 4,268 times greater CR than if SPIRAL, using 9B 'D' for SCM, and 20,152 times greater if 'D' = 42.5B.

Keep in mind the limited visible span along the electromagnetic spectrum. So if the entire visible spectrum is so limited, how much more is the redshifted part of that spectrum.[214]

---

[213] https://arxiv.org/pdf/astro-ph/0012376v1.pdf Hubble survey on Hubble Constant

[214] https://science.nasa.gov/ems/09_visiblelight & en.wikipedia.org/wiki/Electromagnetic_spectrum

Each part of the spectrum is a spectrum. So from mid point of the normal light spectrum, until the start of the redshifted spectrum, there is plenty of spectrum to mask the nearest departure point of visible light here & now subjugated by CE being as near as 5778 LY rather than the rounded 1M LY where detectable CR starts.

Other factors we have touched on like change in density, cosmological blue-shift offset, .. also factor into the distance of the nearest stellar object whose light was subjected to CE and whose light has detectable CR due to that CE.

Velocity and time of course being important to determine an unknown distance. If we know the distance and either velocity or time we can solve for the unknown variable, that may be vital in other hypotheses.

With no change in density factor to help offset, as SCM uses the current consensus assumption of stellar formation 'out there' via gravitational aggregation, so gaining density overall by stellar maturity.

In contrast to SPIRAL's hyper-dense proto-galactic formation, which somewhat offsets the rapid increase in distance w/ CR.

The X (horizontal) axis: Distance 'D' = Light Year (LY) recession distance between us and the stellar object, due to Cosmic Expansion (CE).

Use galactic centers for the average CR of each galaxy, as near side CBO, and enhanced CR on the far side, in any applicable galaxy. Those whose visible light was subject to galactic recession from us due to CE.

From about where the stellar objects are now, whose visible light we see here & now, whose light has been subjected to cosmic expansion on departure, and/or in transit to us.

If SPIRAL the distance to where stellar objects are now, after adjusting for it's orbital velocity and trajectory, are relatively near where they were by the end of the cosmic inflation CI-CE expansion event.

If SCM up to 4B LY, is the maximum distance to where the most distant visible objects were when the light we see here now departed them.

So 2B LY (the mean average starting distance and light departure point of visible stars if SCM-LCDM.

If SCM-LCDM: The light departed up to 4B LY distant, & traveled 13B (rounded) LY to arrive, as subjected by up to 13B (rounded) LY of cosmic expansion, at an average expansion rate of about $13/4 = 3.25$ times light speed (c) with the most distant visible stellar objects here /now being up to 46.5B LY distant. 46.5 - 4B start = 42.5 divided by 3.25 (c) = current max distance after 13B years at 3.25 (c) if SCM-LCDM,

The Y axis: is the cosmological redshift (CR) %. Staring where the nearest object is whose light arrives here now was ever exposed to cosmic expansion, and ending by where the most distant visible (here & now) objects are projected to be by the respective model, now. Overall the greater the distance the more the CR.

Velocity = Speed in relation to c (light speed) = Distance of CE subjugation divided by time of CE.

As the cosmic inflation expansion event lasted up to a day if SPIRAL. So the gross recession speed of a stellar object/galaxy that ended up 1B LY distant was 365(c) during the cosmic inflation expansion event. Of course the way cosmic expansion is understood, is it is not the object that is moving.. but the fabric of space.

Change in Density = Current density dived by density when the light departed it. = LY Distance to where the object was at the end of cosmic inflation expansion (CI) divided by the LY distance to where the visible light from it here & now departed it.

That value is 5778 LY to date and increasing 1 LY per year based on the observations, explanations and corroborating evidence in SPIRAL cosmological model.

So an object at 2B LY /5778 was 346,000 times as dense when the light we see here now departed it, 5778 LY away, 5778 years ago, during the CI-CE epoch, early in the history of the universe.

The radius of the 'Inner Universe' (IU) as used in SPIRAL, is the Light Year (LY) distance to the nearest departure point of any visible starlight, reaching us here & now, ever subjected to Cosmic Expansion (CE).

Be it on departure or in transit. Including not just any light that has been subjected to CE such as that with cosmological redshift (CR), but also to any that has been subjected to cosmological blue-shift (CB).

If SPIRAL we predict it to be a minimum 5778 LY distance ' I ' to the nearest stellar object whose visible light here now was subjected to CE on departure or in transit. 5778 LY to date, add one LY per year to the distance minimum prediction.

As light is limited to 'c' standard light speed, ' I ' caps the number of years elapsed subsequent to CI.

If SPIRAL there is no ongoing cosmological expansion or contraction, So stellar objects are relatively close to where they would be by the end of Cosmic Inflation (CI) Expansion (after taking their orbital velocity into account).

So all visible starlight here now that is from stellar objects beyond distance ' I ' departed those stellar objects when they were at the distance 'I'.

This is similar to CME where the CME radiation, from all the various directions it departed from, reaching here now, traveled for the LY distance = to the years elapsed it took to get here. So if SCM and it has been 13B years, 13B LY years. If 5,778 years, 5,778 LY.

So with SPIRAL the greater the distance over 'I' by the end of CI-CE the denser when the light departed it. A visible stellar object 2B LY away by the end of CI, so was 5778 LY distant when the light we see now from it departed it, the object was not going to reach it's mature density till the end of the CI epoch.

So 2B divided by 5778 = 3,461,000 times as dense. If the object ended 1B LY distant, the density was 1.73M greater, then once the universe was at it's mature size and density, by the end of the CI-CE event, which approximates it's current size.

So 1M LY should correspond to 173 times denser on the light departure. At 57,780 LY 10x. So, exhibit less CR if subjected to CE, than had the stellar object already been at mature post CI density on it's visible light's here & now, departure.

So the increase in density on departure, that increases with the stellar object distance by the end of CI-CE density, helps mitigate the increase of CR with distance.

Without the density offset, one might predict the span of LY distance of stellar objects of which exhibit prevalent increase of CR with distance, we are within the visible spectrum thereof to be vastly below the 13-4=9B LY claimed by SCM.

The visible spectrum being just a small portion of the entire spectrum, so visible CR spectrum being just a portion of the whole visible spectrum.

If so SPIRAL and not Hubble/LCDM best describe the measurements of the natural observations of the increase of CR with distance.

At 9B LY 'D' for SCM is 4,268 times more CR at 100% CR vs SPIRAL with 'D' at 2B LY and 'T' = 24 hrs.

2,134 if 'T' in SPIRAL is 12 hrs, 8,536 if 48 hrs.

20,000 if LY 'D' in SCM is 46.5-4 = 42.5B.

We hope we provided a reasonable equation that best reflects the measurement of CR if SPIRAL and have compared it to what one should predict if SCM.

We also may have exposed the Hubble equation as bending the hypothesis - that CR represents ongoing cosmic expansion, to fit the observations, rather than the equation that best reflects SCM and the observations.

The Copernican principle:

'is one of the primary pillars of the science of astronomy. It says simply that we do not occupy a privileged location in the Universe. As you might guess, the principle comes out of the Copernican view of the Solar System, that the Earth is not the center of the Solar System, the sun is. In other words, we are not the center of the Universe, but rather we are just a small part in the larger whole.

The Copernican principle is one of the primary philosophical underpinnings of the science of astronomy. It is not physical law in and of itself, but it does require us to use an abundance of caution around any hypothesis that requires us to live in a special place of the Universe.' [215]

---

[215] http://blog.professorastronomy.com/2009/05/copernican-principle.html?spref=tw It hijacks Copernicus view of a solar centric universe.

The cosmological principle:

'The idea that the universe should be uniform (homogeneous and isotropic) over very large scales was introduced as the "cosmological principle" by Arthur Milne in 1933. Not long before that, it had been argued by some astronomers that the universe consisted of just our galaxy, and the centre of the Milky Way would have been the centre of the universe. Hubble put an end to that debate in 1924 when he showed that other galaxies exist outside our own.

Despite the discovery of a great deal of structure in the distribution of the galaxies, most cosmologists still hold to the cosmological principle either for philosophical reasons or because it is a useful working hypothesis that no observation has yet contradicted.

Nevertheless, our view of the universe is limited by the speed of light and the finite time since the Big Bang. The observable part is very large, but it is probably very small compared to the whole universe, which may even be infinite.

We have no way of knowing what the shape of the universe is beyond the observable horizon, and no way of knowing whether the cosmological principle has any validity on the largest distance scales possible.[216]

[216]www.math.ucr.edu/home/baez/physics/Relativity/GR/centre.html

If SPIRAL one might predict:

- Prevalent cosmological redshift (CR) of Distant Starlight
- The increase in CR with stellar object distance.
- No CE subjugation within SPIRAL LY radius 'i'.
- Radius 'I' increases of 1 per annum.
- Each viewpoint in the universe has it's respective radius 'I'.
- CR offset due to CB (near side cosmic blue-shift offset).
- Black-Holes by Galactic centers no longer hyper-dense.
- Blue-stars on the near side arms of local Spiral Galaxies beyond 6k LY but not so distant that the CB can't offset the increasing CR, so overall CB dissipates, with distance.
- MVP - We have the optimal view in the entire universe.
- The visible universe approximates the entire universe.
- No OCE, so missing dark energy is not predicted.
- The minimum distance to visible CR departure point.
- The max distance to visible light's departure point.
- The departure point distance of that visible CR light.
- Visible CR departure distance increases 1LY annually.
- Crisp starlight/dark night sky, Olbers' Paradox solved.
- Hyper-dense Proto-Galactic formation by CI solves galactic rotation problem, w/o 'missing' dark matter.
- Magnetic repulsion, Black-hole illusion, CB offset, ..
- Flatness, despite the universe is a sphere w/ us by the approx. center and no OCE. SCM there's no center, but not eternal /infinite, so there are edges beyond which no OCE. So annual dissipation of CMB at 1 LY vs 3.27 LY where OCE, so if SCM checkerboard CMB.
- CMB near black-body whereas SCM-LCDM with OCE only outside gravitational bonded sections would leave more of a checkerboard with a wider spectrum, with 46.5-4=42.5B LY radius worth of dissipation to the edge of the visible universe and a minimum of 13B LY of dissipation if in theory stayed w/in gravitational bound sections. The larger the gravitational bound spans claimed, the bigger the problem for SCM-LCDM.

- Moon is of the Earth, but not as a result of a collision.
  SPIRAL maybe the highest probability explanation of:
- Cosmological redshift (CR) of distant starlight.
- The rate of increase in CR with distance.
- The cause of cosmic expansion (CE).
- Evidence against ongoing cosmic expansion (OCE).
- Why missing dark energy may be non-existent - no OCE.
- Why missing dark matter may be non- existent. The gravity was attracted to where the regular matter was, during a/the hyper-dense phase, that ended relatively early in history, thousands, not billions, of years ago (YA).
- The Galactic rotation issue. Hyper-dense proto-galactic and stellar formation preceded cosmic expansion that ended by the end of literal day four, w/ the universe at mature size and density. Thousands, not billions, of YA.
- Olbers' Paradox. Stellar objects have been emitting their light for thousands, not billions of years.
- The uniform aging of distant stellar objects.
- Blue-stars in outer reaches of 'local' spiral Galaxies
- Blue star 'blue stragglers'.
- Cosmic Blue-shift offset, how if rotating away still blue.
- 'Black-holes' by Galactic centers. No longer hyper-dense as where the matter of those galaxies emanated from..
- No CR of light from stellar objects under xk LY distant..
- That most (all?) stellar formation was early in history. Unless you can prove a stellar object formed in the past 5k years including light travel time, it may be 'all' not most!
- The formation of heavy elements. HTP heat-time-pressure
- The unexpected brightness .. of distant galaxies. See from when but 5780 LY distant to date. So the light was emitted from them when those distant galaxies were denser.
- CMB 'axis of evil' Geo-centric universe alignment.
  SPIRAL align as well or better with 'Galaxy Mergers', 'synchronized radiation', 'ejection pairs' /groups, 'central pinch', 'axis aligned to our line of sight', Lithium levels, and much more

SCM needs fudge on the fudge (dark energy and dark matter fudge). (+ see "Pearlman 'No Fudge' Cosmology" 1/14/2026)

Sept. 25, 2019 / 25 Kislev 5779

'Local observations (that assume ongoing cosmic expansion) indicate universe expanding faster, so the universe is younger than expected. Large Magellanic Cloud Cepheid Standards Provide a 1% Foundation for the Determination of the Hubble Constant and Stronger Evidence for Physics Beyond Lambda-CDM'.[217]

'A new feature in the dark sector of the Universe appears increasingly necessary to explain the present difference in views of expansion from the beginning to the present. '..

'Less predictable but highly sought are contributions from gravitational wave sources as standard sirens (Schutz 1986; Abbott et al. 2017; Chen et al. 2018).

Improvements in parallaxes from future Gaia data releases are also expected to continue to increase the precision of the distance ladder in the near term.'..

Try option five! SPIRAL is obvious for those willing to look outside the deep-time dependent consensus box, All four options considered assume ongoing cosmic expansion. If none, they fail.

If SPIRAL there's no ongoing cosmic expansion! No need for 'new physics' as pretty basic science that can add up and cross check.[218]

'One Number Shows Something Is Fundamentally Wrong with Our Conception of the Universe'.[219]

---

[217] https://arxiv.org/abs/1903.07603   Adam G. Riess, 27 March, 2019
    Magellanic Cloud satellite galaxies of our Milky Way 170 to 210k LY
[218] www.forbes.com/sites/startswithabang/2019/09/13/how-fast-is-the-universe-expanding-incompatible-answers-point-to-new-physics/? Ethan Siegel PhD
[219] www.livescience.com/hubble-constant-discrepancy-explained.html

'Have we been wrong about the age of our universe all along?[220]

Per SPIRAL more apt if we tweak Wendy Freedman quote to 'big bang an expansion (instead of explosion) of, not an explosion into. space' '9% faster than previously thought.'[221]
If Dark Energy and Dark Matter exist/ed.

If isotropic, just as much here as anywhere. So the trace levels alleged by SCM here, too weak to be detected with existing tech., are just as weak out there. While just overwhelmed by the local gravity. Yet at 68% & 27% cumulatively account for 95% of all matter.

If not isotropic, when did it bunch up? And if it did bunch up, why is the CMB still near blackbody? Why the near uniform increase in CR with distance in all directions?

If Dark Matter and Dark Energy isotropic, how is/was planetary formation accretion of ordinary matter via gravity viable?

If Dark Matter (the 27%) evenly distributed, why any net effect on the gravity and the 5% ordinary matter? If it is pulling an even amount from all directions. So not a solution to galactic rotation and other issues.

If Dark Energy (the 68%) the alleged cause of CE is isotropic, than it dissipates/ed along with Cosmic Expansion. If dissipating, it means CE would be slowing, not speeding up, over time. In contrast to the observations.

---

[220] www.thenational.ae/uae/science/have-we-been-wrong-about-the-age-of-our-universe-all-along-1.894032

[221] www.nbcnews.com/mach/science/universe-may-be-billion-years-younger-we-thought-scientists-are-ncna1005541

SPIRAL's hyper-dense proto-galactic formation preceding CE.. requiring just 5%, the known ordinary matter, and no ongoing CE, is a clear internally consistent, with all the empirical observations, explanation and solution.

'The devil is in the details' it is easy to not be falsified if ambiguous. Try to pin down the LY distance to the nearest departure point of visible light here and now, that was ever subjected to cosmic expansion. Any CR light by definition has been subjected thereto.

While SCM hasn't committed to the shape of the 'flat on the surface' universe, it is not eternal/infinite, so there are edges. So the assumed 'Cosmological Principle' distributions over far distances, wouldn't apply past any edge, so a notion 'the universe looks the same wherever you are' is false.

Things don't all add up and cross check under SCM? No need for 'new physics', use basic physics in light of SPIRAL cosmological redshift hypothesis and model.

SPIRAL's **'Dayenu'**

(a play on the popular Passover Psalm and song 'Dayenu' (it would have been enough..).

Had we just found in SPIRAL:

The entire universe approximates the visible universe - Dayenu.

An Earth-Sun Ecliptic centric Spherical Universe - Dayenu

Universe gravitationally bound w/in 4/365(5780) of history- //

Electro-Magnetic repulsion CE force till equilibrium w/gravity- //

The year age 'clock' of the universe = the 'IU' LY radius - Dayenu

The 'OU' outer universe we see from when compressed... -Dayenu

The radius of the entire universe of 1B+/- LY -   Dayenu

Stable, on the first crest, steady-state oscillation universe - //

Past CE from 'macro' galactic, to 'micro' stellar, level - Dayenu

'Inner' (local), vs 'outer', universe from every viewpoint - Dayenu

No ongoing hyper-density at Galactic centers -   Dayenu

'No 'Dark Energy' required as no ongoing comic expansion - //

No 'Dark Matter' required 'GRIP' on Galactic Rotation - Dayenu

Galactic Rotation due to primordial electro-magnetic fields - //

So over 99%, not 5%, of all matter.., is 'normal' matter - Dayenu

SETI Sacked -   Dayenu

Olbers' Paradox – 'Tackled' -   Dayenu

CMB 'Axis of Evil Justice' fits -   Dayenu

CoMBO cosmic inflation 50M:1 parsimony advantage - Dayenu

We see CMB and the entire 'OU' from LY radius 'IU' - Dayenu

No light departed w/in 'IU' radius of any point subjected to CE- //

SPIRAL's 'MVP' our Vastly Preferred View Exhibits A-D - //

Blitz Parallax, Gravity & The Cosmic Distance Ladder - Dayenu

SPIRAL's 'SNAP' Earth Formation Hypothesis -   Dayenu

'Draw Play' lunar & original Continent, formation Hypotheses - //

Pearlman vs Hubble 'Exhibit CE', CR & CE equations - Dayenu

SPIRAL's SAFTEY hypothesis on 'Flatness' -   Dayenu

SPIRAL's 'SOD' on distant object past interactions - Dayenu

SPIRAL's 'GRAB' on Gravitational Bonding -   Dayenu!

Yet, Baruch Hashem, we help determine all of these are higher probability description/explanations of the universe.

Baruch Hashem. SPIRAL resolves a lot of issues on the '..Age, Formation and Structure of the Universe' that are (assuming continue to hold up as valid and the highest probability explanations of the empirical observations, ie the strongest science) worth fistfuls of Nobel prizes:

- 'SPI- (hyper-dense proto galactic formation)   1
- RAL' (cosmological redshift attests that proto-stellar formation preceded cosmic inflation expansion)   1
- 'MVP' (our view is vastly preferred vs any distant view point) hypothesis   1
- 'HTP ' (Galactic & stellar formation hypothesis) 1
- 'Cosmic Blue-shift Offset' hypothesis   1
- 'Black-Hole Illusion Resolution' hypothesis   1
- 'Electro-Magnetic Repulsion' with no Dark Energy 1
- 'CoMBO' 50M:1 CMB.. parsimony advantage   1
- 'GRIP' Galactic Rotation without Dark Matter.   1
- 'Blitz' on the Cosmic Distance Ladder   1
- 'SAFTEY' on the 'Flatness Problem'.   1
- 'SOD' on distant object past interactions   1
- 'GRAB' on Gravitational Bonding'   1
- 'SETI Sacked' save this world as no alt.   1

Parsimony: A small fraction of size & mass accounts for all. We don't even know consensus is possible. As if it could be as consensus, it could certainly be SPIRAL. For example, something with a million has 100. Something with 100 may not have 1M.

SPIRAL explains more, is a far lessor claim, and assumes less. For example, consensus assumes the missing dark energy and dark matter exist.

In science, the greater the claim, the greater the burden of proof. So consensus has a vaster burden of proof. Much of SPIRAL is self-evident. As SPIRAL would predict, rather than just react to, the prevalent CR of distant starlight, the overall increase of that CR w/ distance...

SPIRAL on Keating ten point Big Bang Cosmology checklist:
Dated Dec. 2020. Point 8 edied with new insight August 2022

1. Know your enemy: Learn the mathematical model that undergirds the big bang theory.

    rmp- SPIRAL start at a hyper-dense size. Hyper-dense proto-galactic formation PRIOR to hyper-cosmic expansion. Cosmic expansion, and most stellar formation, end early in history. See 'Pearlman vs Hubble' for the math.

    SPIRAL predicts (vs SCM reacts to) the Prevalent Cosmological Redshift (CR) of distant starlight and overall increase of that CR w/ distance. SPIRAL's 'Cosmological Blue-shift Offset' may account for the Hubble Tension, a tell something off w/ consensus ongoing cosmic expansion explanation of CR.

    SPIRAL we see objects from when they were up to 4B,vs SCM-LCDM's 46.5-4= up to 42.5B LY, closer.

    SPIRAL entire universe approximates the visible universe with estimated radius of 2B LY (4B LY max). SCM-LCDM visible universe 46.5B LY radius, w/ entire universe 500 x the volume of the visible universe. So SCM CMB temp. about 5M times hotter vs if SPIRAL, when both same volume, up to that of a sphere with a radius of 2B LY.

2. Prepare your rival: Your theory of the universe needs to be precisely formulated, ready for anyone to use to predict data.

rmp- SPIRAL = SPI-RAL.

SPI = hyper-dense proto-galactic (Stellar) formation Preceded cosmic Inflation expansion.

RAL = cosmological Redshifted Light, Attests to SPI, as those stellar objects, were subjected to cosmic expansion, during the hyper cosmic expansion epoch.

See 'Pearlman vs Hubble' for equations and 'SPIRAL Predictions' for predictions of CMB temp. distance from nearest departure point of any light ever subjected to any cosmic expansion, and many more predictions made by/if SPIRAL, many of which are self-evident, already confirmed, and /or waiting to be confirmed.

3. Olbers's paradox: Explain why the sky is dark at night.

rmp- Hard to top SPIRAL's explanation thereof or come up with a better design to create such a crisp night sky view.

First: Thousands not billions, of years lapsed, subsequent to stellar formation and cosmic inflation. So no time for light from every object to reach far. Light speed being limited to standard speed of light.

Second: Highly collimated light trails if SPIRAL. As over 99% of stellar objects (all over 5781 LY) in the visible universe, light therefrom visible here and now, departed during hyper-cosmic expansion, when the universe and stellar objects were denser.

SPIRAL predicts as few as 5781 LY to date of normal light (that was never subjugated to any cosmic expansion), emanating from visible and still existing stellar object. Except any stellar objects under 5781 years old within 5781 LY years.

Also see SPIRAL's 'MVP' hypothesis how/why we have a vastly preferred view of the visible, which approximates the entire, universe, over and above any distant vantage point.

So SPIRAL predicts all distant viewpoints beyond our SPIRAL LY radius 'I' observe vastly fewer galaxies beyond their respective radius 'I' than we can. So overall as we approach radius 'I' the darker the night sky gets.

New Horizons spacecraft may have already confirmed this where from it's view some places ten times darker than our view. Reference SPIRAL's 'Olbers' Paradox Tackled' in the book.

Update as of July 07, 2023 Tammuz 18, 5783:
Pearlman YeC 'PVoLT' Olbers' Paradox resolution formula. To measure any model's compatibility with the observed night sky in comparison with SPIRAL cosmological model:
P(L)(T)/V where:
P = Proximity. The presumed size is a function of the proximity of the object. Objects with the same angular diameter of the apparent surface area are as many times (x) larger as they're more distant, so x distant as larger. Example: sun & moon angular diameter (witness a Solar-eclipse), sun 400 rounded x more distant & large.
L = Luminosity per unit of surface area. All else being = the more distant an object the smaller it looks, and less illumination, here. Number of stellar objects assume proportional to volume, over large distances. As more a bigger claim, so larger burden of proof.
T = time span in years light has been emitting from stellar objects.
V = Volume of the visible universe. Volume of a sphere = $(4/3)\P r^3$.
SPIRAL light saturation:
V = of a sphere of 1B LY (rounded) radius(r). 4.19×10^{27}
L = 590B proportional to volume and 'number' if SCM at 100trillion stellar objects. So 0.0059 volume of a 1B LY radius sphere if SPIRAL (entire universe gravitational bound so assume proportional distribution as 'local' region within radius I where SPIRAL and SCM agree on number, size and density of stellar objects) compared to 13B radius if SCM (assumes Cosmological Principal so proportional distribution over large distances).
P = one. Use one for SPIRAL as the baseline to compare competing hypotheses to.
T = 5,783 yrs. (yrs. = to predicted value of SPIRAL LY radius I) to date + 1 per annum. L(P)(T)/V = SPIRAL = 8.13×10^{13}

SCM-LCDM light saturation:

V= a sphere with r 46.5BLY 1.15 x 10^34. L = 100trillion. P = 13 (Proximity factor at light departure 13B vs SPIRAL r 1B). T = 13B yrs. L(P)(T)/V = 4.04 x 10^ -8 = 0.0000000404 if SCM.

Pearlman P-VoLT on Olbers' :

Result: 4.04x10^-8 / 8.13x10^-13 = 49,792.53

SPIRAL has 49.8k less light saturation, so a 49,792.53 advantage on explaining the empirical cosmological observations in regard to Olbers' Paradox vs the current consensus champion competing hypothesis SCM-LCDM.

Both competing models agree: Most stellar formation was relatively early in the history of the universe. That (past) cosmic expansion caused a lot of light to be beyond the spectrum visible to the human eye. To be conservative we round down 13.8B to 13B years for SCM, allowing 800M years less luminosity than the model's estimated age of the universe. Despite JWT findings that indicate earlier galactic formation than predicted under SCM, See 'JWT data aligns best within SPIRAL'.

SPIRAL finds the universe at mature size and density in gravitational bound equilibrium, by the end of 4/365.25 (SPIRAL LY radius i) a fraction of history, by when luminosity up and running. See SPIRAL's 'HTP' hypothesis. At that brief transition from the hyper-dense start, size is made up for by the light intensity.

If our universe eternal would mean bright like day night sky. All else being equal, like the number, size, luminosity and location of stellar objects in the visible universe: - The larger the space V the darker the night sky. - The longer the span T the brighter the night sky.

SPIRAL light year (LY) radius i = distance to nearest departure point of light arriving here and now at 'c' standard light speed, that's ever been subjected to any cosmic expansion. Over 99% (all but 20 out of 100B +/-) galaxies we see from when receding from us, so when being subjugated to cosmic expansion. Any amount of cosmological redshift in a stellar object by definition means it'd been subjugated to cosmic expansion.

4. Redshift and distance: Explain why the light we receive from galaxies is almost always redshifted, and why redshift increases proportionally to distance.

rmp- SPI-RAL where we actually would predict, rather than just react to, the prevalent Cosmological Redshift (CR) of distant starlight and overall increase of that CR with distance.

While SPIRAL predicts the past receding, and overall increase in the rate of that recession with distance. SPIRAL also predicts near-side 'Cosmological blue-shift offset'.

Now the metric expansion due to, and during that, hyper cosmic expansion epoch of space, that ended relatively early in the history of the entire physical universe as a whole, and hyper-dense proto-Galaxies, and individual stellar objects within those galaxies, was not uniform.

As even the rate a single stellar object was subjected to cosmic expansion during that epoch, could vary, based on it's unique position, composition, density, gravitational and electro-magnetic fields influence.

Micro quasar SS-433 particle jets are an example of magnetic field repulsion influence diminishing with distance.

Reference SPIRAL's 'Pearlman vs Hubble' and 'Magnetic-Repulsion' and 'Cosmological Blue-shift Offset' hypotheses.

5. Fading light: Explain why supernovae fade more slowly, and why galaxies are dimmer, when their light is more redshifted.

Rmp- If SPIRAL all cosmologically Redshifted stellar objects, (which by definition have had their light subjected to cosmic expansion), and everything else that exited at the time at their current LY distance, of the light we see now from them and beyond, was subjected to the hyper cosmic expansion epoch, that was relatively early in the history of the universe. SPIRAL predicts by the end of 4/365(5781) a fraction of history, after which there has been no ongoing cosmic expansion.

The nearest stellar object ever subjected to cosmic expansion being at least 5,781 LY to date. Now as their visible light here and now was departing them they were receding from us. Overall the more distant, the faster it was receding. Factor in variable for near-side 'cosmic blue-shift offset' of proto-galactic expansion velocity, that may have varied by Galaxy.

The resting distance of a red-shifted stellar object at the end of a 24 hour hyper-cosmic expansion and time taken for each one second of light emitted therefrom to reach and pass us. 20k LY /86.4k = __84 days_. 200K LY / 86.4k= _2.3 years__. 2M LY /86.4k = _23.15 years_. 2B LY / 86.4k = 23,148 years_.

Thus the more distant, the greater the CR, the dimmer, (as not just more distant, but also lower frequency and longer wavelength), and the slower the fade.

Reference SPIRAL 'Blitz' on the cosmic distance ladder and Cepheid Variables and 'Cosmic Blue-shift Offset' hypotheses.

6. Cosmic microwave background: Explain why the CMB has a near perfect blackbody spectrum at 2.725 K, and why the temperature of that spectrum increases with redshift.

rmp- SPIRAL good with near black-body CMB. SPIRAL even aligns better than SCM w/ CMB 'axis of evil'. SPIRAL may explain why the temp. spectrum increase with CR, as all light arriving here and now that was ever subjected to CR departed when the universe was denser, If the universe was denser the CMB temp was higher. Overall the more distant, the more dense the universe, galaxies and individual stellar objects, when the light reaching us here and now departed the stellar objects.

7. Forest: Explain where the Lyman alpha forest comes from, why it is thicker at higher redshift, and why it is correlated with the positions of galaxies.

rmp- Attests to SPIRAL's hyper-dense proto-galactic formation? Determining the Nature of Late Gunn–Peterson Troughs with Galaxy Surveys https://doi.org/10.3847/1538-4357/aac2d6 via @IOPscience

8. Cosmic oven: Explain why the universe (especially the parts untouched by stars) is 75% hydrogen, about 24% helium, and about one nucleus in 40,000 deuterium. Protip Link: https://alterbbn.hepforge.org/
rmp- SPIRAL: Early element, primordial gas cloud, black-hole, quasar, and Stellar formation, Prior to cosmic Inflation.[222]
Aligns as well or better with abundance and distribution of light elements 'cooked' and cooled relatively early in history. The universe attains mature size and density within 4 days: 4/365(SPIRAL LY radius 'i') a fraction of history. The entire universe approximates the visible universe, which approximates a sphere with a 1B LY radius.

---

[222] Latif, M.A. et al. **Turbulent** cold flows gave birth to the first quasars. Nature 2022

Hydrogen, Helium,, great design & creation, employed to get from the hot and dense proto-universe, to where we are here & now.

How hot the universe was when smaller, depends on the given current CMB temperature, and how large is the universe now. Both competing hypotheses agree the universe started in a minuscule size.

Until larger than SPIRAL, when SCM- LCDM the same net area, it's 137.7M times hotter. A sphere radius 1B vs one 46.5B (1370).

SCM-LCDM consensus the entire universe is at least 250 times as large as the visible universe, estimated at LY radius of 46.5B. So at 250x, if the elements formed at a certain temp., they formed 25.1M times earlier (when the universe was 25.1M times smaller) if SPIRAL vs if SCM, where they already formed relatively early in history.

Now 1370 times is where timing of the CMB photon emission epoch after 1 day if SPIRAL, equates to 375k years if SCM.

Start 137.7M times cooler if SPIRAL, than if the universe is 137.7M times as large, as per SCM-LCDM. As CMB cools with cosmic expansion. Double the size, half the temp. So 137.7M cooler if SPIRAL than if SCM, at any like size along the way.

SPIRAL radius 'i' = the nearest light LY departure point of any light arriving here and now at 'c' standard light speed, ever subjugated to cosmic expansion, that ends after the hyper-cosmic expansion epoch early in history. With expansion vs gravitational forces in equilibrium.

Avg. CMB temp. at x size = Current avg. CMB temp / (current area entire universe/ x area universe). www.lenntech.com/periodic/elements/h.htm We are still learning.[223]

---

[223] https://phys.org/news/2022-08-nitrogen-extremely-unusual-high-pressure.html
'HTP' Proto-galactic & element formation could've been nearly instantaneous.

9. Do better: Explain or eliminate dark energy and dark matter, decipher satellite galaxies, solve the lithium problem, analyze the absence of antimatter, and unveil the beginning (or not) of the universe.

Rmp- SPIRAL is the best for the missing Dark Energy and the missing Dark Matter. As neither is required or predicted if SPIRAL. (For example if SPIRAL there has been NO ongoing cosmic expansion subsequent post hyper-cosmic expansion). Yet without them the competing SCM-LCDM, that require both, is falsified.

Reference SPIRAL's 'GRIP' on galactic rotation (that obviates the need for Dark Matter) and 'Magnetic Repulsion' on cosmic expansion, hypotheses.

'solve the lithium problem' rmp- SPIRAL 2.5M times earlier, than already relatively early SCM-LCDM, on turbulent primordial gas cloud, and element, formation. Favors 'large-scale inhomogeneities in cosmic density'.[224]

'absence of anti-matter' rmp- In scriptural testimony via Moses One super-natural creator, who in 10 utterances (equations) creates, then organizes, all the energy and matter in the physical universe, along with the laws/forces of nature, over the course of creation week one.

Think of the utterances as the vibrating strands of energy that all matter breaks down to. Withdraw that which sustains the vibration, the physical universe reverts to nothing. See Pearlman YeC vol. I Torah & Science alignment framework.

---

[224] https://ui.adsabs.harvard.edu/abs/2017nova.pres.1997K/abstract & Wikipedia 'On possible variation in the cosmological BARYON FRACTION' Astrophysical Journal, Vol. 716, Num. 2 Gilbert P. Holder et al 2010

10. Publish: Present your theory - principles, equations, and predictions - concisely and clearly to the light of expert scrutiny. Engage scientists through the right channels, be patient, and strike a blow for the revolution! (Just be sure to give Luke and Geraint at least some of the credit.).

SPIRAL was first published direct to book format in 2013. Early peer-review scrutiny by John G. Hartnett PhD, feed back from several scientists on researchgate.net, researching peer-reviewed articles, and familiarizing myself with the basic science, and issues where the empirical observations and basic science needed consideration and reconciliation.

The likes of which Dr. Keating and guests often discuss, that at times I comment on and share. Some being included in this checklist, the result being many updates and advances in science, within SPIRAL.

Extra credit: leave a comment with the analogous checklist, but for Theories of Everything!

If SPIRAL over 99% of matter is normal matter. No need for a singularity to start or within black holes Reference SPIRAL's RambaN 'mustard seed' area start reference and 'Black Hole Illusion Resolution' hypothesis

'Big Bang' checklist from the interview on 'Cosmic Revolutionary's Handbook: How to BEAT Big Bang Cosmology' with Geraint Lewis & Luke Barnes, by Professor Brian Keating, you-tube 'Theories Of Everything' series, link at: https://youtu.be/3Wg1mRPRg0c . Please watch, enjoy, like, share and subscribe.

End SPIRAL on Keating Big Bang Checklist.

Oct. 8, 2021 / 2 Cheshvan 5782, Yahrzeit Akiva b. Pinchas Vogel

Pearlman SPIRAL's **'SAFTEY'** hypothesis! helps advance science, this time with the 'Flatness Problem'.[225]

SAFTEY stands for: SPIRAL cosmological model's Answer to the Flatness problem issue Empirical observations Testifies to a Young universe.

In the first Friedmann equation[226] we find the size of the universe – scale factor represented by 'a' is a variable. The larger the universe the tighter the already tight range, that the empirical observations have any chance of being explained by SCM-LCDM current consensus big bang cosmology champion.

So all else being equal, SPIRAL has an advantage over current SCM of the area of a sphere with a 13B LY rounded down, divided by that of a sphere with a radius of 5782 LY. A Quintilian parsimony advantage for SPIRAL, assuming a straight line increase in the 'Flatness Problem' with area, that Robert Dicke helped bring to light.

CMB visible here and now departed from a uniform distance. From the far edge of the sphere that is the visible universe. So our view-point is opposite mid-base, of any two CMB hotspots. So an Isosceles triangle. At any point in time there is only one base distance actuality, between two CMB hot spots. So the triangle angular measurements are based on distance assumptions that depend on the model.

If SCM 13B (rounded) LY uniform sides of triangle. The angle from here is going to be much smaller, thus conclude 'flat' universe. With the entire universe radius a volume over 125M times visible universe.

In addition SPIRAL's much cooler start and 'Jiffy Pop' trillion(s):1 advantage on initial magnetic repulsion zones.

---

[225] https://en.wikipedia.org/wiki/Flatness_problem
[226] https://en.wikipedia.org/wiki/Friedmann_equations

VS larger angle if SPIRAL 5,782 LY distance of the sides of triangle, that jibes w/ spherical.

As if SCM, the most distant visible / detectable stellar objects, light reaching us now, departed about 4B LY distant, 13B years ago, and due to an assumption of ongoing cosmic expansion, have traveled about 13B LY to reach us.

Whereas in SPIRAL, we conclude no ongoing cosmic expansion subsequent to a hyper cosmic expansion epoch that ended by 4/365(5782) a fraction of history to date.

We are seeing the most distant stellar objects over 5782 LY from when they were 5782 LY to date (add one LY per annum) distant. The more distant the stellar object the denser it was when the light we detect from it here and now departed it. See 'Pearlman vs Hubble'.

Another tremendous advantage SPIRAL has over SCM on the Flatness problem, is if SPIRAL, hyper-dense proto-galactic formation was PRIOR to hyper cosmic inflation, when all the matter and energy was in close proximity. So no need to walk the fine-tuning tightrope of galactic formation via accretion, after The Big Bang and after cosmic inflation.

Now just like with CMB where both competing hypotheses (SPIRAL and SCM) agree we see here and now from a uniform distance (5782 LY vs 13B LY) so not much depth. So too SPIRAL holds distant starlight depth is up to 5782, LY vs SCM 13B LY.. So if the visible universe looks 'flatter' then one would expect under SCM, the actuality being SPIRAL might reconcile the observations.

Also the more distant a stellar object, the smaller it should look, all else being =. So too the denser a stellar object, the smaller it's area vs when it is less dense, all else being =.

So with SPIRAL where we observe stellar objects over 5782 LY when they were denser, the more distant, the denser, the earlier on hyper cosmic expansion day four, we see them from.

See SPIRAL on the classic leavening raisin dough analogy, with the caveat the galaxies themselves began hyper dense. So were expanding to reach mature size and density by the end of day four. So we are seeing the universe when mostly denser, from a mostly uniform departure point, is a lot less depth so we would predict, should appear 'flatter'.

Also factor in the effects of gravity on light (so I assume CMB cosmic microwave background radiation too) . A straight line being the, just as closest distance between two points. So light a 'c' standard light speed and light speed limit, that is bent, or tugged at, should take longer to reach here than if t a straight line. Here for simplification, we assume straight line and speed of 'c'. For in depth discussion on gravity on light see the experts at the 'Dynamics of Gravitational Fields' lab I'm part of at Researchgate.net.

I posted a bit about this on (association of orthodox Jewish scientists 'AOJS' about Sept. 2021 after Dr. Brian Keating noted 'the Flatness Problem increases with distance'. That turned on a light in my head that SPIRAL provides a higher probability explanation, just as SPIRAL does with an over 150T:1 parsimony advantage over the competing hypothesis SCM-LCDM cosmological model.

If cosmic inflation's a solution, SPIRAL & SCM share. Yet that may just compound the fine-tuning challenge on those in denial of Intelligent Design of the universe.

In conclusion, SPIRAL 'SAFTEY' has a fourfold solution to the 'Flatness Problem' over the competing hypothesis, the current consensus champion SCM-LCDM:

Volume: SPIRAL has a Quintilian 1.136568979E19:1 parsimony advantage over SCM. The issue increases with area. Front row vs nosebleed stadium seating, on steroids. It'd be like trying to view a movie here positioning a standard IMAX projector beyond Pluto. When the longest screen to projector is about 100 feet, so starts to get blurry within a mile.[227]

Tuning: Proto-Galactic formation was PRIOR to hyper-cosmic expansion if SPIRAL. So no need for very low probability fine tuning, required after cosmic expansion.

Depth: The first 5782 LY should be about the same, under both models. If SCM there should be about 13B LY of Depth to the visible universe, whereas if SPIRAL the entire universe attained mature size and density by the end of day 4 w/ a radius of 4B LY max we see all beyond 5782 LY when 5782 LY. So 5782 LY of Depth is visible to date.

Density: the fourth advantage SPIRAL has over SCM, as the more distant the stellar objects, the denser they were when the light we see from them departed them. As they crossed that distance earlier on cosmic expansion day four.

So SPIRAL would predict, and aligns with, the observations that the 'Flatness Problem' increases with distance. 'Flatness' helps corroborate SPIRAL and falsify SCM-LCDM!

If SPIRAL at over 5782 LY the closer the object, the closer to mature size and density, the later on hyper cosmic expansion day 4 we see the stellar objects from. By the end of day 4 the universe attained mature size and density. The entire universe approximating the approximate sphere of the visible universe, with a radius estimated at 2B LY, 4B LY max.

So now 99%+ we see now from residue of day 4 on a 'flat-screen' @ 5782 LY distance. So over time we will see 1 more LY of 'normal' light annually till within 4B years no flat-screen. Like light subjected to past cosmic expansion, if SPIRAL

---

[227] See SPIRAL on Olbers' Paradox and the Cosmic Distance Ladder

SPIRAL cosmological model **'SOD'** hypothesis
on distant objects that interacted in the past.
Dated Jan. 07, 2022 / 05 Shevat, 5782

SOD stands for:

SPIRAL's Hyper-Dense Proto-Galactic Formation is the -

Optimal explanation for -

Distant Structures that interacted when in closer proximity.

In SPIRAL, we conclude no ongoing cosmic expansion subsequent to a hyper cosmic expansion epoch that ended by 4/365(5782) a fraction of history. We're seeing all distant stellar objects over 5782 LY from when they were closer: 5782 LY to date (add one LY per annum) distant.

The more distant the stellar object the denser it was when the light we detect from it here and now departed it. See 'Pearlman vs Hubble'.

If SPIRAL, hyper-dense proto-galactic formation was PRIOR to hyper cosmic inflation, when all the matter and energy was in close proximity. Not galactic formation via accretion after The Big Bang and Cosmic Inflation.[228]

The more distant a stellar object, the smaller it should look, all else being =. So too the denser a stellar object, the smaller it's area vs when it is less dense, all else being =.

If SPIRAL we observe stellar objects over 5782 LY when they were closer and denser. Overall, the more distant they ended up by the end of day four, the more dense, and the earlier on 'cosmic inflation day four', we see them from.

SPIRAL on the classic leavening raisin dough analogy: galaxies themselves began hyper-dense. If SPIRAL not just gravity, but Electro-magnetic radiation played a large role in galactic formation.[229]

---

[228] SPIRAL 'HTP' hypothesis on galactic and stellar formation. Where we find conditions for near instantaneous stellar formation.
[229] SPIRAL ' Magnetic Repulsion' hypothesis on cosmic expansion

Either way we factor in the effects of gravity on light. For in depth discussion on gravity and light see 'Dynamics of Gravitational Fields' lab I'm part of at Researchgate.net.

The current consensus requires deep-time. Be it distant 'mice galaxies', or other such affected structures we have already mentioned in SPIRAL, or in our own galaxy.[230]

Yet with consensus assumption of ongoing cosmic expansion all that time, the default norm would be these objects would not have gotten closer to each other but been distancing further apart. If not, why not more opaque?

Under consensus our attribute observations require imaginative speculation like 'by no longer existing galaxies'.

If SPIRAL closer proximity, and more dense, so stronger gravitational interaction plus electromagnetic interaction, during hyper-cosmic expansion.[231]

SPIRAL Parsimony advantage 45.76 Trillion : 1 = 365/4(1B) (290M/5782). Based on stellar objects interacted over the course of 1B yrs. Vs 4 days, and 290M vs 5,782 LY travel time and distance. See NGC 4676 'Mice Galaxies'.

SCM-LCDM fights uphill against entropy, over time elapsed and distances traveled. So a crippled alternative to the competing hypothesis SPIRAL cosmological model.

In conclusion, SPIRAL's grip tightens as the highest probability explanation of our empirical cosmological observations, now including of distant stellar objects past interaction.

---

[230] Structures Discovered In The Milky Way Suggest Our Galaxy Looks Different https://youtu.be/ggPLVNXydfA Anton Petrov 12/22/2021
www.nasa.gov 'When_Galaxies_Collide'. Both ignore Olbers' Paradox.
[231] SPIRAL's Cosmic Distance Ladder & 'GRIP' hypothesis on galactic rotation.

Pearlman YeC SPIRAL 'GRAB'
Dated: March 13, 2022 / 10 Adar II, 5782.

If SPIRAL a **Gra**vitationally **B**ound (GB) universe relatively early in history. When repulsive and gravitational forces reached equilibrium. When the entire universe, which approximates the visible universe (2B LY estimated radius) attained mature size and density. At the end of the hyper cosmic 'inflation' epoch, that ended the number of years ago = to 'I'. 'I' = the LY distance to the nearest departure point of light ever subjugated to any cosmic expansion (CE), reaching us here and now.

While most consensus based estimates for 'I' are between 15k to 5M LY, once we consider all the science, like SPIRAL 'Cosmic Blue-Shift Offset' and the spectrum between mid-yellow and red, and TD Chronology, one can predict 'I' = 5,782 to date and that epoch ended 4/365(5782) a fraction into the history of the universe and time. So by the end of cosmic inflation day four. There being no credible, internally consistent model, less than that.

Some considerations on the Local vs Universal GB dispute:

- Checkerboard vs Near Black-body CMB. If GB is localized (by Galaxies) but not universal, the CMB would become cooler where any ongoing cosmic expansion, relative to GB regions, on it's transit here from the LY voyage it has undertaken = to the number of years elapsed, from about the time of the end of cosmic expansion. Yet we do not find checkerboard CMB. So near black-body CMB testifies not only to a hyper-cosmic expansion epoch early in history, but universal, not multi-regional GB.
- The one region we can examine best, that within 5782 LY to date, is GB without dispute. Which would be true if the entire universe is GB (and only maybe true if not).

- Even if one were to assume the current consensus of ongoing cosmic expansion, why is there uniform no GB rather than a spectrum, within radius 'I' ? For more than average bonding where more gravity, less where less gravity.
- Why did large objects that generate more gravity than in our region, not in our region, show evidence of moving apart due to CE. Cosmological Redshift, by definition, is evidence of being subjugated to CE. Past if SPIRAL vs ongoing if SCM-LCDM.
- Distant objects that interacted in the past not only indicate that localized GB did not prevent them from moving apart in the past, but that CE was prior to the onset of GB, and SPIRAL (hyper-dense proto-galactic formation prior to cosmic inflation) is the strongest (highest probability) scientific explanation.
- If CE is so weak as to not disturb regional GB, now that the visible universe has a radius of 46.5B LY per SCM consensus champion, then how could CE have had a chance anywhere, when all much denser, (in a 4B LY radius 13B YA)?
- Dark Energy is required and predicted to exist if ongoing CE is true. It's still 'dark' meaning it is not directly visible, which would be the case if it does not exist. SPIRAL successfully predicts why we still see residue of past, not ongoing, CE. See SPIRAL's 'Pearlman vs Hubble'.

I hope one or more point above helps tip the scale in favor of SPIRAL's GraB vs the competing consensus hypothesis, on the disputed universal vs regional, GB. Please enjoy, recommend and share this research.

Also comment with any additional observations on this specific issue here, as peer-review scrutiny helped make Pearlman SPIRAL the (know it or not) leading cosmological model.

Considerations on what amount of subjugation to cosmic expansion and change over time one would predict if SPIRAL vs if SCM. Leavening raisin dough is the classic analogy, where the raisins stay put in their respective place, yet distance from each other due to cosmic expansion. In SPIRAL hyper-dense proto-galactic formation the 'raisins' also expand. Assume a steady expansion rate, where the radius of the visible universe doubled every 1-3 hours (and still doubles every 389M years if SCM).

Observations perceived as an accelerating expansion rate may be due to SPIRAL describing the actuality. Where we see all the more distant stellar objects from when they were at SPIRAL LY radius 'I', 'I' years ago, during Hyper cosmic expansion day four. The more distant over radius 'I' the stellar object, the earlier during that epoch the photons, we see it from here and now, departed it.

SPIRAL radius 'I' = the nearest departure point of any photons reaching us here and now that were ever subjected to any cosmic expansion.

All (both competing hypothesis) agree the local region is 'gravitationally bound' for at least 5,783 LY. If SPIRAL the entire, approximates the visible, universe and attained 'gravitation bound' equilibrium by the end of day 4. All agree the prevalent cosmological redshift of distant starlight and increase of that CR over distance, visible here and now, is due to cosmic expansion at a minimum of 5,783 years in the past.

- To simplify the math, start at 1 LY radius size.
- The visible universe light year (LY) radius approximates: 46.5B if SCM. 1B if SPIRAL.
- SCM visible universe increase in LY radius during the last doubling 46.5B /2 = 23.25B. 13.8B years / 35.x doublings to get to 46.5B LY radius) = every 389M years rounded. 23.25B LY/389M = 60LY current annual radius increase if SCM. Now we see from photons that departed stellar objects when x LY distant, that traveled x+y LY, due to y LY subjugation of cosmic expansion.
- SPIRAL the visible universe attained mature size and density by the end of day four.

Rounded: LY radius doubled 30 times to get to 1B LY radius. The last radius doubling 1B/2= 500MLY. The radius doubled every hour (3 max depending on when first at 1LY). The more distant the stellar object, the earlier we see it from. The more cosmic expansion it's been subjected to. The more distant from center 'raisins' recede faster.

- So reached radius 'i' sooner. Use the adjusted LY distance proportional to where an object is now. We view from when it was at LY radius 'I' vs radius I-92 LY, 92 years ago. Take change in density and light departure point into account.

Over radius 'i' adjust to the relative distance in the SCM consensus cosmic distance ladder, to the end of the visible universe. So a galaxy estimated at 9.3B LY if SCM, if SPIRAL estimate at 200M LY now. Taking into account independent metrics, such as SPIRAL radius 'i', where the object was when the light we see here and now departed it.

So if we have size and luminosity Tolman test data from 1930 and data of the same x LY distant object 92 years later.

If SCM one would expect about 92(60)(x/46.5BLY) of distancing due to cosmic expansion.

If SPIRAL one would expect about 92LY/x increase in the amount of cosmic expansion the light we see here and now has been subjected to. (take Cosmic Blueshift Offset and relative distance into account). While there's no ongoing cosmic expansion, we see from 92 years later and light-years more distant, when receding faster. The more distant the object now, the greater the increase in velocity.

For example, a stellar object 1% the distance to the 1B LY end of the universe = 10M LY. A sphere that has a radius of 10M LY that began at 1LY would have doubled over 23 times. Unlike SCM where we are seeing the light that departed from the stellar objects relative position if SPIRAL if over radius 'i" we see from when the light departed at radius 'I'.

So distant stellar objects expanded past what radius 'I' was 92 years ago. In seconds, minutes or hours max, prior to expanding past radius 'I' today which is 92 LY more distant. Due to the metric expansion of space during the 'day four hyper cosmic expansion epoch. That began no earlier than day one. 92 LY/10M LY is the increase in the amount of cosmic expansion the object had been subjected to, based on when we see it now vs 92 years ago.

The Tolman test measures for change in cosmological redshift, that by definition is due to the photons / light being subjugated to cosmic expansion.

The Tolman Test also takes into account the distance to the object. As all else being equal the more distant an object the smaller it appears. For example, the moon and sun appear to an observer here as about the same size (think of an eclipse). Yet the sun is about 400 times larger and more distant. The moon would appear 400x smaller if it was the distance of the sun.

If SPIRAL 'radius i' recedes one LY per annum. To an extent the reduction in the size it appears due to this distancing, is offset by the transition during day 4 of the galaxies from hyper-dense to mature size and density. See SPIRAL's 'HTP' hypothesis on galactic and stellar formation.

If SCM expect a non-gravitational bound in relation to us stellar object to distance $92(62.8)(x/46.5BLY)$ over the course of 92 years due to subjugation to assumed ongoing cosmic expansion. See SPIRAL's 'Pearlman vs. Hubble'.

Here we assume a uniform doubling in size cosmic expansion rate of the approximate sphere, that is the visible universe, that we (Earth-sun ecliptic) are the approximate center of.

Ninety-two years of Tolman test data align best with SPIRAL (past cosmic expansion) and SCM (ongoing cosmic expansion), and not with a static universe.

Cosmic Expansion steadily doubling rate SPIRAL & SCM

Spans (start at 1)	LY	Radius Growth	Average
1	2	1	2
2	4	2	2
3	8	4	2.67
4	16	8	4
5	32	16	6.4
6	64	32	10.6
7	128	64	18.3
8	256	128	32
9	512	256	56.9
10	1,024	512	102.4 ..
11	2,048	1,024	
12	4,096	2,048	
13	8,192	4,096	
14	16,384	8,192	
15	32,768	16,384	
16	65,536	32,768	
17	131,072	65,536	
18	262,144	131,072	
19	524,288	262,144	
20	1,048,576	524,288	52,428.8
21	2,097,152	1,048,576	
22	4,194,304	2,097,152	
23	8,388,608	4,194,304	
24	16,777,216	8,388,608	
25	33,554,432	16,777,216	
26	67,108,864	33,554,432	
27	134,217,728	67,108,864	
28	268.435,456	134,217,728	
29	536,870,912	268.435,456	
30	1,073,741,824	536,870,912	35.8M

31   2.1B (If SPIRAL 4B LY radius max.).
32   4.3B. 33 8.6B. 34   17.2B.   35 34.4B. 36 68.7B so
35.x spans = 46.5B LY radius. 46.5B / 35.x = 1.3B average radius
increase per doubling.

Each radius doubling, each part of the whole doubles. So cosmic expansion is cumulative subjugation to 1M light years thereof, be it over 1 hour or 1k years, should net the same result. Each radius 1x1x1=1 doubling = 2x2x2= 8 fold volume increase. A sphere being pi/4 of a cube spanning 2x it's radius.

If SCM: 13.8B years/35.x spans as now 46.5B = 388.7M. Last doubling 23.25B / 388.7M = 60 LY radius, 60(8)= 480 volume, current annual increase, of the visible universe.

If SPIRAL: If visible universe 1LY radius after:
6 hrs. day one: 90 hrs. / 30 spans = doubled every 3 hrs.
12 hrs. day two: 60 hrs. /30 = doubled every 2 hours.
18 hrs. day three: 30 hrs. /30 = doubled every 1 hour.
Per Pearlman vs Hubble: no ongoing cosmic expansion subsequent to day end four, 4/365 (SPIRAL LY radius 'I') a fraction into history. Radius 'I' = the nearest departure point of any light visible here and now, that's ever been subjected to any comic expansion. Predict radius 'I' pans out to be 6k rounded to date. 5,783 to be more precise. Some rounding above.

Dated: Oct. 4, 2022 / 9 Tishrei, 5,783 anno-mundi
Pearlman YeC for the alignment of:
 Torah testimony, science and ancient civilization.
Volume II: 'SPIRAL cosmological redshift hypothesis and model'
www.amazon.com/dp/B07DP4TBZ5

'Hamotzi Mechavero Alav Haraayah'[232]

If we're talking about science and fair consideration, the greater the claim, the greater the <u>burden of proof</u>.

So parsimony is an advantage in weighing the probability of competing hypotheses.

Assume both competing hypotheses have an equal amount of evidence, ie the same empirical observations.

SPIRAL vs SCM-LCDM

1:20 Ratio of matter. SPIRAL 'normal' 5% is 99% of all matter.[233]

1:2.4M 5781 / 13.8B yr. old universe. Time fights uphill vs Entropy.[234]

1:12,568 volume SCM 46.5B vs SPIRAL 2BLY radius visible universe.[235]

1:125M If SCM the entire universe volume a min. 125Mx the observable universe.[236]

So 20 x 2.4M x 12,568 x 125M = x and LCDM is x times greater a claim. So SPIRAL x times a more reasonable claim than LCDM.

X =9.426e+19 = 94 Quintilian + 260 quadrillion.

---

[232] T' Bava Kamma 46a The burden of proof is on..

[233] SCM: 'Normal' matter is just 5% of all matter.
SPIRAL: normal matter approximates all matter and no requirement for the missing Dark Matter, and Dark Energy, required by LCDM to have a chance of being a viable hypothesis. 5% is 1/20 of 100%.

[234] SCM: 13.8 Billion year age of the universe based on deep-time dependent assumptions such as the Copernican Principle of Modern Cosmology and Hubble Expansion explanation for cosmological redshift..

[235] SCM: the most distant visible stellar objects we see now departed a maximum of 4B LY distant. Due to assumed ongoing cosmic expansion (CE) their light traveled 13.x B LY for the same # of years to reach here and are now up 46.5B LY distant. 46.5-4=42.5 LY /13 billion years = an average distancing of 3.27 standard light speed.
SPIRAL: CE ends at the end of Hyper cosmic expansion day four, so after 4/365(5781) a fraction of history, so for now we assume a minimum radius of 400k LY, a max radius of the 4B, so 2B LY a happy medium.

[236] Fermi Lab Don Linclon PhD Dec. 22, 2021 based on CMB hot-spot angular measurements using SCM distance assumptions.
If SPIRAL: the visible universe approximates the entire universe.

For % multiply by 100. All else being equal had SPIRAL been just 10% more reasonable than LCDM it should become the new standard.

That it is 9.426 sextillion percent more reasonable, the only reason for it not to be the new standard is because it hasn't been fairly considered and disseminated yet.

Now aside from that staggering statistical reason, there are many other material reasons SPIRAL is a higher probability explanation of the cosmological observations than LCDM.

- SPIRAL would predict, rather than just react to, the prevalent Cosmological Redshift of distant starlight and overall increase of that CR w/ distance.
- SPIRAL would predict, rather than just react to, the appearance of 'Black-holes' at galactic centers.
- SPIRAL the strongest resolution of Olbers' Paradox.
- SPIRAL helps explain Blue-shift in galaxies up to _200M?_ LY.
- SPIRAL aligns better w/ CMB 'Axis of Evil'.
- SPIRAL doesn't require Dark Matter & Dark Energy.
- SPIRAL predicts light departed w/in LY radius 'I' not subjected to cosmic expansion.

Every view point in the universe sees 'normal' light from stellar objects within their radius 'I'. The entire universe reached 'gravitationally bound' equilibrium by the end of 'hyper cosmic expansion day four', when it attained mature size and density.

From objects beyond 'I' we see their light trails that departed them when they were at that radius 'I' LY distance, which is equal to the years elapsed after the end of hyper cosmic expansion. The same LY distance, and years ago, CMB departed from that is reaching here now.

James Web Space Telescope (**JWST**) observations favor SPIRAL
Updated Aug. 12, 2022 / 15 Av, 5782

Earlier galactic formation than one would predict if SCM-LCDM consensus, is predicted if SPIRAL This mirrors the failure of NDT Darwinism doctrine consensus to predict the functional complexity that appears abruptly & relatively earlier in the fossil record. The stronger science there being within ID & YeC.[237] Here too Pearlman YeC best explains the JWST observations:
- Galactic formation relatively earlier in history...[238]
- Size of distant stellar objects.[239]
- Past galactic interaction that shaped many (most?) of the galaxies.[240]
- Distance to, and time lapse of, the observations, including 'Tidal Disruption Events', X-Ray Binary, and GRB's.[241]

---

[237] On Intelligent Design (ID) see Discovery Institute.
The strongest ID is w/in YeC. See Pearlman YeC, ICR.org and AIG.com.
[238] "Panic! At the Disks: Optical Observations at Z>3 JWST of SMACS 0723 Field (arxiv.org) [2207.09428] +
"A very early onset of massive galaxy formation. Labbé, Ivo et al.(2022) +
"Two Remarkably Luminous Galaxy Candidates Z11-13 Revealed by JWST."
Naidu, Rohan P. et al.(2022).
[239] JWT 'too big' (not if light departed from SPIRAL radius 'i'). at 6:50 on
https://youtu.be/SF3wF2IfkRc Anton Petrov
[240] Ancient Collision Far Away Created The Most Powerful Explosion
https://youtu.be/OBMh-mMs3G8 Anton Petrov
JWT Reveals Cartwheel (+ ring, elliptical, spiral,..) Galaxy Secrets
https://youtu.be/AEOdKKGHcOs Anton Petrov
Summary of Major JWT Galactic Discoveries That Nobody Expected
https://youtu.be/5YGbqlFb3yg Anton Petrov
[241] Horesh,A.,Cenko,S.B.,& Arcavi,I.(2021). Delayed radio flares from a tidal disruption event. *Nature Astronomy*,1-7.

'My take on @NASAWebb results thus far - were already seeing strong evidence that galaxy formation is occurring much quicker and earlier than we supposed. This has been a recurring theme with HST, Spitzer, Keck, etc, all finding the same.'[242]

In SPIRAL radius 'I' is defined as the nearest light year (LY) departure point of any light arriving here and now that has ever been subjected to any cosmic expansion. We see stellar objects beyond radius 'I' from when they were at that LY distance, as the universe went from hyper dense to mature size and density, on 'hyper cosmic expansion day four'. Objects radius 'I' and closer we see from their 'normal light' never subjected to cosmic expansion. So take the years = to their LY light departure distance to reach here.

SPIRAL hypothesis: Cosmological Redshift Attests to Hyper-dense Proto-Galactic formation Prior to Hyper-cosmic expansion. Where the entire universe approximates the visible universe. Attaining mature size and density by 4/365(SPIRAL LY radius i) a fraction into history, at a radius of about 2B LY. The time of observation here vs time elapsed at departure, is a function of cosmic expansion subjugation. SPIRAL includes 'Cosmic Blue-shift Offset', 'Blackhole illusion resolution', 'GRIP', 'HTP', 'MVP' and many additional sub-hypotheses.

---

[242] Professor Christopher Conselice, July 20, 2022

'Lyman-Break, Spectroscopy, aligns with SPIRAL' Jan. 02, 2023

JADES - JWST Advanced Deep Extra-Galactic Survey finds relatively early Re-ionization and stellar formation. 'Thousands not billions' of years ago as SPI-RAL (Stellar formation PRIOR to cosmic Inflation.) explains. Based on the strongest science (highest probability explanation of empirical observations). Time to know of 'SPIRAL cosmological redshift hypothesis and model'. Hidden in plain sight, just a few steps outside the confirmation bias lined consensus box.

'Determining Distance Through Spectrography.. The Lyman Limit emission is a specific fingerprint .. originates in the Ultraviolet part of the electromagnetic spectrum...' [243]

'Spectroscopy to reveal an incontrovertible distance and redshift for this object.' 'some regions achieving full reionization earlier and others later'. 'waves of re-ionization'..,

'The spectroscopic identification of the Lyman break signature, present and easily visible in all four ultra-distant, JWST-identified galaxies, confirms their redshift and distance.' [244]

Providing 'key constraints on models of galaxy formation and evolution, as the predicted abundance varies greatly when different physical prescriptions for gas cooling & star formation are implemented.[245]

'Photometry and spectroscopy to help ascertain redshift (relative) distance and age. Galaxies w/ photometric and spectroscopic redshifts.'[246]

Actual distance and age are model dependent. Calibrate using the same data & redshift. JWTS photometry and spectroscopy results align as well or better with SPIRAL as with the competing hypotheses, current consensus SCM-LCDM. Expect SPIRAL to replace as the new standard in cosmology, after study, fair consideration & dissemination.[247]

---

[243] https://futurism.com/determining-distance-through-the-lyman-limit

[244] JWST breaks Hubble's record! link.medium.com/VykaFw9xUvb Ethan Siegel

[245] Bright z-9 Galaxies in Parallel: https://arxiv.org/abs/2205.12980

[246] Two Remarkably Luminous Galaxy Candidates at z - 10–12 Revealed by JWST https://doi.org/10.3847/2041-8213/ac9b22

[247] SPIRAL: 'HTP' on galactic formation, 'Magnetic Repulsion' and 'Pearlman vs Hubble' on Cosmic

SPIRAL hyper-dense proto-galactic formation preceded
' hyper cosmic-expansion **day four**' Aug. 12, 2022 / 15 Av, 5782

What time of day four do we see which stellar objects from, when the light we see here and now, departed from them, when they were crossing light year (LY) radius 'I', during CI?

Envision the classic leavening raisin dough analogy, but w/ expanding raisins.

Assumes:

- The hyper cosmic-expansion day 4 epoch lasted 24 hours.
- Start day 4 at a 1 LY, ends at a 2 Billion(B) LY, radius.
- The universe doubled in size every x minutes. X=46.5
- 'Day four' begins 'Tuesday eve at 6PM ends Wed. at 6PM.
- The entire universe approximates the visible universe.
- The universe attained mature size & density by end day 4.
- Radius i = 5,782 to date, plus increase of one per annum.

Starting with 1, double 31 times to get to 2B rounded. 24 hours (60 minutes) = 1,440 minutes. 24/31=.774. 774 (60-minute hr.) = 31 sets of 46.45 minutes = 1,440 min. So the universe reached a radius of 5,782 after 12.5 sets. 12.5 sets of 46.5 minutes /60 = 9.7 hours.

So, when the light departed them on day 4, the most distant galaxies in the universe, that we see here & now 5,782 yrs. Later were (rounded):

By 4AM 5,782 LY distant. The radius of the entire universe.
At 9AM the radius of the visible (entire) universe 200k LY
At 12 Noon - 20M LY. 4:26 PM a radius of 500M LY.
At 5:13 PM universe at a radius of 1B LY .
By Wed. 6PM the universe at mature size & density 2B LY radius.

Use the above to derive how many hours, days, weeks, months or years, is the observation span for what transpired in how many seconds, as the universe transitioned from miniscule to a radius of 1 LY to mature size and density at radius 2B LY, by the end of four days.

Like the moon appears the same size as the 400 times more distant and larger sun. So too the same object x more distant appears x times smaller and vice versa. A closed system that increases in volume by Y, decreases in density by Y and vice versa.

Size and Age of the Universe based on CMB temperature:
Dated: January 4, 2026 / 15 Tevet, 5786

Here we compare the two competing cosmology models Pearlman Cosmology and the current consensus champion SCM on the size and age of the universe based on CMB temperature.[248]

Both models assume a hyper-dense start followed by a hyper cosmic expansion 'inflation' epoch. Conclude: Abrahamic Faith in Mosaic Testimony aligns best with basic physics.

In SPIRAL (hyper-dense proto-galactic formation Precede most cosmic 'Inflation' expansion) cosmological redshift hypothesis and model, we find the entire universe approximates the visible universe. It attained gravitational bound equilibrium early on by the end of 4/365.25(i) a fraction into history, 'i' years before present, by a radius of 1B Light Years (LY) rounded.

'i' = LY distance to the nearest departure point of any light arriving here and now at standard light speed 'c', that was ever subjugated to any cosmic expansion. So a few million cap, but predict it will pan out at the minimum, 5786 to date, once one factors in all the applicable science, in historic context.

All beyond SPIRAL radius 'i' is look-back to 'The First 96 Hours' aka '4-day cosmic inflation'. The more distant past i, the denser the universe, and higher the CMB temperature, at light departure from it, reaching us here and now, at 'c'.

SCM-LCDM (SCM) assumes ongoing 'Hubble Expansion' over 13B+ years and the volume of the entire universe is at least 250 times the observable universe, now at radius 46.5B LY. So if SCM the CMB temp. should be at least 250 times higher when the entire universe condensed into the size of the observable universe. Start size: between the radius of a photon and 1 millimeter.

---

[248]Re .. temperature evolution law of the CMB with Gaussian processes -
https://arxiv.org/abs/2505.24543_ Felipe Avila, et al

Try these values for SPIRAL that assume SCM-LCDM has at least has some of the relative timing, temp. and size right:

CMB **Kelvin** Temperature

Light Year radius observable universe

Time post hyper-dense start

SCM vs SPIRAL		Event	SCM vs SPIRAL		SCM vs SPIRAL	
2.73	2.73	Observable Universe	46.5 B	1B	13.8B years	96 hours to now
60	3M-3k?	Re-ionization/proto-galactic formation:	2B	1-1k	150M-1B yrs.	x hours
3k	3M-3k?	'CMB Last Scattering':	42.3 M	$1 - 560^{249}$ $- 912k^{250}$	380k yrs.	x hours
20k	5M-20k?	'Radiation = matter energy'	?	0.x	10k yrs.	x minutes
10M	10M-1M?	'CMB spectrum fixed'	?	0.0x	1 month	x minutes
1B	1B-1M	'Nuclei form':	?		3 minutes	3 minutes
10B	10B-10M	'Neutrinos Decouple':	?		1 second	1 second
10T	1T-1B	'Protons & Neutrons form':	0.004 LY?	x miles	10 micro-seconds	10 micro-seconds
10Q	1Q-1T	'Grand Unification Era':	3m	.x meters(m)	1 Planck Time	1 Planck Time

SCM-LCDM: 'The universe doubles in size every 10-34 s. Inflation stops at around 10-32 s, by which time the universe has increased in size by a factor of 1050. This is equivalent to an object the size of a proton swelling to 1019 light years across!'[251]

---

[249] www.researchgate.net/publication/382524758_SPIRAL_universe_size_at_d ecoupling_CMB_calibrated 560 before factoring in SCM, the entire, is 250+x the observable, universe. So if 560x, then 1.0.

[250] SCM radius of 42.3M to 46.5B doubling 10.1 times = the 10.1 of going from radius 911,806 to 1B LY.

[251] https://cms.cern/physics/story-of-the-universe

Keep in mind there are many definitions and understandings of cosmic 'inflation'. So SPIRAL's increasing the duration up to 96 hours and reducing the estimated volume of the entire universe today to a sphere of radius 1B LY rounded (as no subsequent cosmic expansion), allows for a much lower starting temperature.[252]

SPIRAL's 'Jiffy Pop' of primordial electro-magnetic fields approximating the number of existing galaxies may also help explain.

One variable is the hyper-dense starting size. For this paper assume from between the radius of a photon and millimeter (near Ramban's 'mustard seed' hyper dense start description/analogy).[253]

Can both model's be compatible with a 2.725K near blackbody CMB temperature distribution? How to test and repeat?

If SCM-LCDM: With the radius now is 46.5B LY and 2.725K. A 3K Temperature and 42.3M LY radius at the surface of the last scattering. Both 1/1100 of that now.

So for most of history averages a straight function of the radius (if the radius halves the temp halves, if the radius double the temp. doubles). If always so 2.725(46.5B) K = x the temp at a radius of 1.0 LY. Considering CMB Temperature as more a function of volume, adjusting the start size and/or inverse square curve, can help close the gap of SPIRAL vs SCM.

The conclusions could be model dependent. A change in volume can also effect a change in the temperature. As can other dynamics, like energy transference.

A steady doubling rate would mean a progressively faster expansion. The more distant the stellar object, the greater the velocity. Offsetting that the inverse square rule, Electrostatic Force weakens with distance. So a progressively slower rate of doubling. So a more even expansion speed. By 96 hours gravity and electromagnetic repulsion are in gravitational bound equilibrium. End cosmic inflation and expansion.

Consider cosmic inflation/expansion using the classic leavening raisin dough analogy. If SPIRAL the raisins expanded too, as hyper-dense proto-galactic formation. With the entire universe approximating the visible universe. 4/365.25(i) a fraction into history. 'i' years before present. At a radius of 1B LY rounded.

[252]The First Three Seconds on CI-CE https://arxiv.org/pdf/2006.16182 Jan.'21
The First Three Minutes / Beyond .. Steven Weinberg, Aug. '93 / Jan.'80
[253]The Ramban (1194-1270), Elisha and a Pot of Oil: How the Universe Makes Itself - Jewish Action Gerald Schroeder (2009)

Equations can help describe the relationship between time, temperature and size (radius size and/or volume, some use redshift 'Z'). SPIRAL finds the LY radius of the observable universe is 1/46.5 of that claimed if SCM. That = 1/100,546 a fraction of the volume.

Conclusion: Either way, based on radius or volume or a dynamic in-between not nearly as hot to start, allowing a spherical universe, decoupling, neutral gas / atoms within seconds, minutes or hours, and allows for proto-galactic formation starting within 24 hours, 72 max, at a radius closer to 1LY. With gravitational bound equilibrium by a 1B LY rounded radius, by the end of 96 hours. So a viable alternative to SCM starting at a radius of 1B LY after 200M years. Also proto-galctic formation could precede the timing of the last scattering.

Modern Cosmology is 'cumulative' the recipe for the multi-generational reinforced confirmation bias. That precludes consideration of higher probability models (like SPIRAL) that reject the Copernican Principle, Cosmological Principle, and/or ongoing cosmic expansion.[254]

Some context:

Volume of a sphere of LY radius 46.5B is 100,546 times that of radius 1LY (= 4.189 cubic LY). Geometric Sequence: Use an 8 fold increase in volume for every doubling of the radius.

Average annual rate of cosmic expansion from the hyper-dense start to date (distance/time elapsed):

SCM 46.5/13.8 = 3.37 times standard light speed 'c'.

SPIRAL 1B/6k = 166,667 times standard light speed.

High speed cosmic 'inflation' expansion as fast or faster doubling of size if SCM.

___ times doubling, to get from radius __ to radius __:

10.15 photon to millimeter.   63.1 millimeter to 1LY.

    31.0 1LY to 1B LY.

SCM: 5.54 1B LY to 46.5B LY.   SCM: 10.1 42.3M LY surface of last scattering to 46.5B LY.

SCM: 2.49 to increase volume 250 times. Doubling 3.0 increases the volume 8x8x8 = 512 times.

SPIRAL: if start radius 1MM: doubling 94.1 times within 96 hours = near hourly on average.

Cite: www.researchgate.net/profile/Roger-Pearlman-Yec/research

---

[254] www.researchgate.net/publication/326427142_Pearlman_vs_Hubble
www.researchgate.net/publication/381802842_SPIRAL_'MVP'_Hypothesis

'Did New Horizons Detect Dark Matter Light? Excess Optical Light Mystery'.[255]

RMP – rather than evidence of Dark Matter, this may be confirmation of SPIRAL. See SPIRAL 'MVP' hypothesis exhibits A-D. Where this may be the light trails from far side, more distant, galaxies that preceded it passing SPIRAL LY radius 'I' on cosmic expansion day four, 4/365(SPIRAL LY radius I) a fraction into history. When the universe transitioned from hyper-dense to mature size and density.

To stick to current consensus is Olbers' Paradox on steroids. Plus inconsistent. Why block all the other light for asserted billions of years but not this light. Why there and not closer to here where easy to test and confirm if the alleged type of Dark Matter is real or not. New Horizons deep-space probe: page 2 of 2

NASA: New Horizons Spacecraft answers 'How Dark is Space'.[256] RMP – New Horizons already far enough to view   the night sky 10 times darker than this view. Helping confirm SPIRAL 'MVP' hypotheses where we (Earth-sun ecliptic) have The most preferred view over and above any distant view point in the universe. So predict a darker night sky (far fewer visible galaxies) from any distant view-point.

Mostly only those galaxies within SPIRAL LY radius 'I' of any distant view point and those galaxies on the far side of it are visible from there. So the vast majority of the entire universe appears without visible stars and galaxies.

---

[255] Anton Petrov https://youtu.be/eb0XZe3k-SM presentation on: Cosmic Optical Background Excess, Dark Matter, and Line-Intensity Mapping. José Luis Bernal, ..Phys. Rev. Nov. 2022

[256] www.nasa.gov/feature/new-horizons-spacecraft-answers-question-how-dark-is-space

So the night sky appears much darker by the time one has a viewpoint past SPIRAL LY radius 'I' (the distance to the nearest light departure point of any light arriving here and now, that's ever been subjected to any cosmic expansion. Predict as low as 5,783 to date, and add 1 per annum.

The main question if SPIRAL is overall at what distance does it start to get darker, away from the focal point of cosmic expansion, that we are in proximity of (if not we would not be able to view the light from distant stellar objects in all the various directions, as if SPIRAL proto-galactic formation was PRIOR to hyper cosmic expansion, that ended 4/365 (SPIRAL LY radius I) a fraction into history, thousands not billions of years ago and a light speed limit of 'c')

Mankind: Special and Central

Aug. 23, 2022 / 26 Av, 5782

Ours is the most preferred view in the entire universe. Vastly better than any distant viewpoint.[257]

The competing hypothesis, consensus champion SCM-LCDM, puts the radius of the visible universe at 46.5B LY and the entire universe at least 200 times larger. So SCM claims the area of the entire universe is 2.5M larger than if SPIRAL.

It turns out galactic formation was earlier than one would predict if SCM, yet as predicted if SPIRAL. If SPIRAL 2.5M times earlier galactic formation, than even the relatively early galactic formation reality, than if SCM consensus.[258]

For argument's sake, assume 1M other civilizations were created out there evenly distributed in the visible universe, even using SPIRAL's smaller 2B LY radius. The odds of contact with complex life out there, that did not come from here (excluding Hashem, angels and other created forces of nature), is zero.

Conclude: It's all for us. The entire universe is for each individual human 'Earthling' to grow our connection with The One designer, creator, sustainer, sovereign of the universe, our One common Father, aka G-d of Abraham, via exercising our free will to be aware, abide by and study our respective covenants of 7-613. The opposite is to opt to live in denial and experience spiritual entropy. Free will to be aware is an amazing design feature. In place since literal day six and still working like new. Ref.: 'Pearlman YeC' alignment of Torah, Science & Ancient Civ.

---

[257] Pearlman YeC SPIRAL 'MVP' hypothesis

[258] "Panic! At the Disks: Optical Observations at Z>3 JWST of SMACS 0723 Field (arxiv.org) [2207.09428] + "A very early onset of massive galaxy formation. Labbé, Ivo et al.(2022) +
"Two Remarkably Luminous Galaxy Candidates Z11-13 Revealed by JWST." Naidu, Rohan P. et al.(2022).

'SETI Sacked' [259]

Dec. 08, 2022 / 14 Kislev, 5783

In Pearlman YeC SPIRAL cosmological redshift hypothesis and model (Hyper-dense proto-galactic formation...) the observable universe we are by the approximate center of approximates the entire universe.

That attained mature size and density, at a radius estimated at 1B Light years (LY) rounded, by the end of day 4/365(SPIRAL LY radius 'I') a fraction into history.

Based on the highest probability pure science, there's almost zero chance life 'evolved' even once (including here) via NDT Darwinism. For NDT to even have a chance of being a viable hypothesis requires hundreds of millions of years of continuous life sustaining conditions, to get us to this point.

'Thousands not Billions' before present. Only the number of years = to SPIRAL LY radius 'i' have elapsed subsequent to the hyper-dense super-hot start of hyper cosmic expansion 'inflation'. Prior to day 5 conditions not yet ready compatible with life.[260]

SPIRAL with it's parsimony advantage of 5.67 Quintilian:1 based on size, density and a straight-line entropy factor, over the competing current consensus champion hypothesis SCM-LCDM (expect SPIRAL to replace as the new standard, after fair study & dissemination). For all practical intents & purposes falsifies all deep-time dependent scientific hypotheses & assumptions.

Radius 'i' defined as the nearest LY departure point of light visible here and now, that's ever been subjected to cosmic expansion. So thousands, not even 5M LY and years ago.

---

[259] Psalm 115:16. Save our world as no alt. civilization. SETI Search for ExtraTerrestrial Intelligence www.history.nasa.gov/seti.html

[260] & Latif, M.A.,et al. www.nature.com/articles/s41586-022-04813-y Turbulent cold flows gave birth to the first quasars. Nature 2022.
& Anton Petrov: 'How Did Massive Quasars Form So Early In the Universe?,.. Turbulent origins not conducive for any life'.

By definition stellar objects with cosmological redshift have been subjugated to cosmic expansion.

It turns out the distance to the nearest stellar object subjugated to any cosmic expansion is closer, than to that of those already exhibiting CR. See SPIRAL's 'Pearlman vs Hubble'.

That leaves ID, and the vastly stronger ID science is within YeC creation science. Now based on what we know from Torah testimony, that alone has stood the test of time for the scope, span and detail, covered therein, 'i' = 6k rounded. 5,783 + one per annum to be more precise, once the inflated history books are corrected and the science catches up. See Pearlman YeC.

There's no motive or reason for advanced organic life to have been created other than here. Assume a communication and travel speed limit of 'c' standard light speed. The odds of contact with complex life out there, that didn't come from here till the end of human of flesh and blood history, is near zero.

Excluding Hashem, angels and other created forces of nature, and perhaps vegetation, as the Earth seeded by day 3. The day the moon parted from the Earth. See SPIRAL 'Draw-Play'.

As the sum of the time elapsed till they send out a signal after day 4, plus time for that signal to arrive, can't exceed SPIRAL radius 'I'. We are at the center of the sphere of SPIRAL with radius 'I' which is a tiny fraction of the entire universe. There is no realistic chance of contact with intelligent extraterrestrial life, for well past a fair expectation for our existence on Earth.

R' Dr. Lamm on the implications if extraterrestrial life ever makes contact with us doesn't consider Pearlman SPIRAL.[261]

With intelligent life, including Angels, including man, created in mature stature by the end of literal day 6.[262] Pearlman YeC Alignment of Torah and Science w/ Ancient Civilization.

---

[261] https://traditiononline.org/the-religious-implications-of-extraterrestrial-life/ Psalm 68:18 + T' AZ 3b: Hashem roams 20k-2k=18k worlds. Either Angels created starting day 2, and/or Pearlman: H' visiting souls in Olam Habba. T' Chagigah 12b. Those not yet born &/or those already passed on.
[262] R' Yehoshua b. Levi Tractate Rosh Hashanah 11a

Abbreviations + some definitions & explanation:

'c'   = Light Speed. The standard speed of light.

CB   = Cosmological Blue-shift offset

CE   = Cosmic Expansion of space (including CI)

CI   = Cosmic Inflation Expansion (if SPIRAL all CE)

CMB = Cosmic Microwave Background Radiation

Cosmological Principle: universe on large scales homogeneous & Isotropic.

Copernican Principle: our location in the universe isn't privileged.

CR   = Cosmological Redshift (due to CE)

GRB   = Gamma Ray Burst

IU = Inner Universe = Visible Light from stellar objects therein was never subjected to Cosmic Expansion (CE) so not CR'ed or CB'ed. LY radius = to year count post CI.

LCDM=SCM Lamda Cold Dark Matter w/Dark Energy & Reg. Matter

LY   = Light Year/s. Distance light travels per year at 'c'.

NASA = National Aeronautics and Space Administration

OCE   = Ongoing Cosmic Expansion / Hubble Flow

OU= Outer Universe Starlight therefrom was subjected to CE, departing when universe still compressed, prior to mature density.

SCM   = Standard Cosmological Model /LCDM

SSSO = Stable Steady State Oscillation

STR = Einstein's Special Theory of Relativity

Visible (Observable) Universe = Sphere of detectable matter,
        Estimated LY radius if: SPIRAL 1B. If SCM 46.5B.
        Earth - sun ecliptic orbit the approximate center of the visible universe. If SPIRAL the visible, approximates the entire, universe.

SPIRAL= SPI-RALL=Predicts CR and it's overall increase w/distance.

SPI = hyper-dense proto-galactic (Stellar) formation Preceded (near all) cosmic-Inflation expansion.

RA = Cosmological Redshift Attests to their (those stellar objects receding from us during cosmic inflation).

LL = Lagging Light (decreased frequency/increased wavelength) as that distancing extends their arrival time-span.

More abbreviations:

AM   = Anno Mundi (year count, yr. one start day six)
B,M,k = B=Billions, M=Millions, k not K = Thousands
DTD   = Deep-Time Dependent Doctrine
ID   = Intelligent Design
RCCF = The Recent Creation Framework six principles
TDC   = Torah Discovery Chronology
YA   = Years Ago
YeC   = Young Earth Creation

SPI-RALL = Hyper-Dense proto-galactic (Stellar) formation Preceded cosmic Inflation expansion. Cosmological Redshift Attests to their Lagging (decreased frequency / Longer (increase) wavelength) Light. So Cosmological Redshift is the result of, and attests to, those stellar objects receding from us during the cosmic inflation expansion epoch. That was relatively early in the history of the physical universe.

'SPIRAL' cosmological redshift hypothesis and model
Edition 770-R: www.amazon.com/dp/B09L3NP43P
Article with DOI, after July 07, '23 edition 770-Q until Jan.19,'26
Cite: Roger M. Pearlman YeC   CC BY 4.0
Title / Description   DOI-Link   Month / Year

New Cirinus Black Hole data in light of SPIRAL. January 2026
DOI:10.13140/RG.2.2.10171.07207
The First 96 Hours - Size and Age of The Universe based on
CMB Temperature. DOI:10.13140/RG.2.2.36546.44482 Dec. '25
Radio Dipole Anomaly resolved! by Pearlman SPIRAL cosmological
redshift hypothesis and model on the cosmic distance ladder.
DOI:10.13140/RG.2.2.16275.34080   November 2025
Quasar Dipole and CMB in light of Pearlman SPIRAL cosmology
DOI:10.13140/RG.2.2.10806.36165   November 2025
Pearlman Cosmology hyper-dense proto-galactic formation helps
explain the age and structure of the universe.
DOI:10.13140/RG.2.2.21890.18881   May 2025
SPIRAL universe size at decoupling CMB calibrated.pdf
DOI:10.13140/RG.2.2.29656.20484   April 2025
Pearlman Cosmology first and best explanation of Oxygen in the
earliest light trails b.pdf DOI:10.13140/RG.2.2.26144.49925 Apr. '25
Feynman's Doubt DOI:10.13140/RG.2.2.35508.36484 Dec. 2024
CMB distribution not 'being 'checkerboard' is explained best by
Pearlman SPIRAL cosmology and discuss the gravitationally
bound region relation to SPIRAL light year radius i.
DOI:10.13140/RG.2.2.29855.98721   August 2024
Pearlman Cosmology on 'Rakiah' in scripture - Metric Expansion of Space.
DOI:10.13140/RG.2.2.30240.49923   March 2025
Pearlman YeC SPIRAL: No Fudge Cosmology   July 2024
10.13140/RG.2.2.32319.52641 & 10.13140/RG.2.2.13069.96480
SPIRAL cosmological model 'MVP' Hypothesis   June 2024
DOI: 10.13140/RG.2.2.14477.35041
JWST Rapid Mass Assembly and Metal Enrichment findings
confirmation of Pearlman YeC SPIRAL cosmological redshift
hypothesis and model   June 2024
10.13140/RG.2.2.20373.13287 & 10.13140/RG.2.2.24054.87367

Table of Contents:

---

[263] Einstein's Doubt, Biggest Blunder and Regret Oct.'20 DOI:10.13140/RG.2.2.14194.72648

[264] Parallax, Gravity and The Cosmic Distance Ladder April '16 10.13140/RG.2.2.12752.93449

[265] 'GRIP' Galactic Rotation without Dark Matter Oct. '20 :10.13140/RG.2.2.11678.14404

[266] Olbers'_Paradox_Pearlman_YeC_evaluator June '23 DOI :10.13140/RG.2.2.29530.72648

[267] SPIRAL's Magnetic Repulsion file 7.46b pdf July '20 DOI:10.13140/RG.2.2.11829.13281

[268] 'Blue-shift Offset' (to Cosmological Red-shift) hypothesis part of 'The Pearlman SPIRAL'
cosmological red-shift hypothesis.. May 2016 DOI :10.13140/RG.2.1.2072.9205

[269] SPIRAL cosmological model's 'Black-hole Illusion Resolution at Galactic Centers'
January 2018 DOI :10.13140/RG.2.2.31968.08966

[270] Pearlman YeC SPIRAL 'Jiffy Pop' 'Electro-'Magnetic Repulsion'
on Cosmic Inflation Expansion   Jan. 2018 DOI:10.13140/RG.2.2.11914.07364

[271] Pearlman 'GRaB' hypothesis March 2022 DOI :10.13140/RG.2.2.23153.75361

[272] Distant Starlight with CREST, CMB comments April'16 :10.13140/RG.2.2.19378.88002

---

[273]Pearlman YeC SPIRAL 'CoMBO' Hypothesis Dec. '21 DOI:10.13140/RG.2.2.30985.04962

[274]SPIRAL vs SCM relative dating based on CMB Aug. '22 : 10.13140/RG.2.2.15438.93760

[275]'MVP' hypothesis .. exhibit A   January 2015 DOI: 10.13140/RG.2.2.33824.99846

[276]'Draw Play Lunar Formation Hypothesis.. April '17 DOI:10.13140/RG.2.2.11075.21289
    SPIRAL's 'Draw Play' info-graphic   June 2017 DOI:10.13140/RG.2.2.15200.07687

[277]'SNAP' Proto-Earth origins hypothesis January 2018 DOI:10.13140/RG.2.2.20224.03841

[278]'Pearlman vs Hubble'   July 2018 DOI:10.13140/RG.2.2.23703.59043

[279]Dayenu Fistful Nobel's Worth scientific advances Aug.'20 10.13140/RG.2.2.10852.39044

[280]Pearlman on the Keating Checklist   Aug. '22 DOI:10.13140/RG.2.2.11676.91528
    Pearlman Lab to test SPIRAL...   May '21 DOI:10.13140/RG.2.2.11787.50728

[281]'SPIRAL 'SAFETY'   Oct. '21 DOI:10.13140/RG.2.2.13061.27367

[282]Pearlman SPIRAL 'SOD'   Jan. 2022 DOI:10.13140/RG.2.2.36817.75360

[283]SPIRAL vs SCM-LCDM on Tolman Test Oct. 2022 DOI:10.13140/RG.2.2.12530.11201

[284]JADES-JWST Lyman-break   Jan. '23 DOI:10.13140/RG.2.2.20079.85924
    and JWST helps corroborate SPIRAL Aug. '22 DOI:10.13140/RG.2.2.15840.56322

[285]SPIRAL hyper-dense proto-galactic formation preceded 'hyper cosmic-expansion day four'
    SPIRAL update 'Cosmic Inflation, Day Four' Aug. '22 DOI:10.13140/RG.2.2.13354.63687

[286]Mankind is The Focal Point of The Universe Aug. '22 DOI:10.13140/RG.2.2.16290.64963
    We are special. SETI Sacked.pdf   Dec. '22 DOI:10.13140/RG.2.2.16675.31520

www.ingramcontent.com/pod-product-compliance
Lightning Source LLC
Chambersburg PA
CBHW051448170526
45166CB00001B/162